Frank Hall Knowlton

Fossil Flora of the Yellowstone National Park

Frank Hall Knowlton

Fossil Flora of the Yellowstone National Park

ISBN/EAN: 9783743348028

Manufactured in Europe, USA, Canada, Australia, Japa

Cover: Foto ©berggeist007 / pixelio.de

Manufactured and distributed by brebook publishing software (www.brebook.com)

Frank Hall Knowlton

Fossil Flora of the Yellowstone National Park

FOSSIL FLORA

OF THE

YELLOWSTONE NATIONAL PARK

BY

FRANK HALL KNOWLTON

EXTRACT FROM "GEOLOGY OF THE YELLOWSTONE NATIONAL
PARK," MONOGRAPH XXXII OF THE UNITED STATES
GEOLOGICAL SURVEY, PART II, CHAPTER XIV

WASHINGTON
GOVERNMENT PRINTING OFFICE
1899

CHAPTER XIV.

FOSSIL FLORA OF THE YELLOWSTONE NATIONAL PARK.

By FRANK HALL KNOWLTON.

HISTORICAL SUMMARY OF WORK ON THE FOSSIL FLORA.

As nearly as I have been able to determine, the first collection of fossil plants made in the Yellowstone National Park was obtained by members of the United States Geological Survey under Dr. F. V. Hayden, in 1871. They were found in two localities, and were recorded by Prof. Leo Lesquereux,[1] as follows: "Divide between the source of Snake River and the southern shore of Yellowstone Lake," and "Near Yellowstone Lake, among basaltic rocks." It has not been possible to rediscover these localities, and several of the species remain unique.

In the following year (1872) the Park was again visited by a party under Dr. Hayden. The members of this party investigated the north-eastern portion of the Park and discovered the rich plant deposits on the Yellowstone River, a short distance below the mouth of Elk Creek. The actual collectors were Messrs. A. C. Peale, Joseph Savage, and O. C. Sloane. The plants represented five species, which were determined by Professor Lesquereux.[2]

The Fossil Forest, that has since become so widely known, was first described by Mr. W. H. Holmes in 1878.[3] He visited and quite thoroughly explored the Fossil Forest and vicinity and made a small collection of plants that were submitted to Professor Lesquereux. Most of these plants were determined to be new to science, but they were neither named nor

---

[1] Ann. Rept. U. S. Geol. and Geog. Surv. Terr. for 1871, pp. 295, 299.
[2] Op. cit., Rept. for 1872, p. 403.
[3] Op. cit., Rept. for 1878, Pt. II, pp. 47-56.

described. Holmes pointed out the fact, since abundantly confirmed, of the succession of forests that have been entombed one above another. His section of Amethyst Mountain shows clearly this remarkable phenomenon.

In October, 1874, Dr. Otto Kuntze, a celebrated German botanist, then on a botanical exploring journey around the world, visited the Park and made some interesting observations on the process of petrifaction of trees now going on in the vicinity of the geysers and hot springs. His paper was not printed, however, until his return to Germany in 1880.[1]

The thorough exploitation of the Park was begun and carried on for several years by the Yellowstone Park Division of the present Geological Survey. In 1883 the work was extended toward the northeastern portion of the Park, and collections of greater or less extent were made at many places. In 1885 the Fossil Forest section was worked out, and large collections were made by Mr. Arnold Hague, Mr. W. H. Weed, Mr. George M. Wright, and Prof. J. P. Iddings.

In the summer of 1887, Prof. Lester F. Ward and I spent about six weeks in the vicinity of the Fossil Forest, making large collections of fossil wood and leaf impressions. The exact localities are enumerated below.

The following season I spent two months in the same area, discovering many new beds of plants and more thoroughly exploring and collecting from beds previously known. These are also recorded in the following list of localities.

LIST OF LOCALITIES AT WHICH FOSSIL PLANTS HAVE BEEN COLLECTED IN THE YELLOWSTONE NATIONAL PARK.

1871.

Divide between the source of Snake River and the southern shore of Yellowstone Lake; Hayden survey. (Not since observed.)

Near Yellowstone Lake, among basaltic rocks; Hayden survey. (Not since found.)

1872.

Yellowstone River below mouth of Elk Creek; A. C. Peale, Joseph Savage, and O. C. Sloane.

1878.

Amethyst Mountain in vicinity of Fossil Forest. W. H. Holmes.

Pro. Ausland. Bot. pp. 504-507; 556-560; 660-672; 681-689.

1883

Andesitic breccia near gulch northwest of peak west of Dunraven; J. P. Iddings, September 12, 1883. (Field Nos. 86, 77.)

Needle Hill near Yanceys; W. H. Weed, October 9, 1883.

Tower Creek; Arnold Hague, September 16, 1883. (Field Nos. 1039, 1034.)

1884

Fossil Forest section, lower stratum; Arnold Hague, September 24, 1884. (No. 1224.)

Fossil Forest section, middle stratum; Arnold Hague, September 24, 1884. (No. 1220.)

Fossil Forest section, upper stratum; Arnold Hague, September 24, 1884. (Nos. 1217, 1218, 1219.)

Sandstone on top of ridge west of Mink Creek; Arnold Hague. (No. 2332.)

1885

Signal Hill; W. H. Weed, September 28, 1885.

Head of Tower Creek; W. H. Weed, September 25, 1885.

East slope of high hill three fourths mile south from Yanceys; George M. Wright, September 4, 1885.

Near top of south wall of canyon of Yellowstone River, about 1 mile up stream from mouth of Hellroaring Creek; George M. Wright, September 9, 1885.

Fossil Forest section, No. 4 of section; W. H. Weed and George M. Wright, September 19, 1885.

Fossil Forest section, No. 15a of section; W. H. Weed and George M. Wright, September 19, 1885.

Fossil Forest section, No. 22a of section; W. H. Weed and George M. Wright, September 20, 1885.

Fossil Forest section, No. 29 of section; W. H. Weed and George M. Wright, September 20, 1885.

Fossil Forest section, No. 26 of section; W. H. Weed and George M. Wright, September 20, 1885.

Top of Mount Everts, west face, nearly opposite bridge over Gardiner River, between Mammoth Hot Springs and Gardiner; George M. Wright, July 7, 1885.

1887

Fossil Forest, bed No. 1, lowest bed about 7,500 feet altitude; Lester F. Ward and F. H. Knowlton, August, 1887.

Fossil Forest, bed No. 2; Lester F. Ward and F. H. Knowlton, August, 1887.

Fossil Forest, bed No. 3, "Magnolia bed," 500 feet above bed No. 1; Lester F. Ward and F. H. Knowlton, August, 1887.

Fossil Forest, bed No. 4, "Aralia bed," 425 feet above bed No. 1; Lester F. Ward and F. H. Knowlton, August 20, 1887.

Fossil Forest, bed No. 5, "Salix bed," about 400 feet above bed No. 1; Lester F. Ward and F. H. Knowlton, August 19, 1887.

Fossil Forest, bed No. 6, "Platanus bed," 425 feet above bed No. 1; Lester F. Ward and F. H. Knowlton, August 19, 1887.

Fossil Forest, bed No. 7, highest bed, 515 feet above bed No. 1; Lester F. Ward and F. H. Knowlton, August 19, 1887.

Hill back of Yanceys, near standing trunks; Lester F. Ward and F. H. Knowlton, August 16, 1887.

Cliff west of Fossil Forest Ridge, near Chalcedony Creek, lowest bed, altitude about 7,500 feet; Lester F. Ward and F. H. Knowlton, August 15, 1887.

Cliff west of Fossil Forest Ridge, upper bed, 250 feet above lower bed; Lester F. Ward and F. H. Knowlton, August 15, 1887.

East end of Fossil Forest Mountain, bed on same horizon as fossil trunks; Lester F. Ward and F. H. Knowlton, August 13 and 22, 1887.

Specimen Ridge, head of Crystal Creek, opposite mouth of Slough Creek, "Platanus bed," altitude about 7,500 feet; Lester F. Ward and F. H. Knowlton, August 24, 1887.

Specimen Ridge, opposite mouth of Slough Creek, "Quercus bed," 100 feet above "Platanus bed," Lester F. Ward and F. H. Knowlton, August 25, 1887.

North of Pinyon Peak, on Wolverine Creek, altitude 7,900 feet; Arnold Hague, August 10, 1887.

1888

Yellowstone River, one-half mile below mouth of Elk Creek, bottom of bluff; F. H. Knowlton, August 29, 1888.

Yellowstone River, one-half mile below mouth of Elk Creek, 30 or 40 feet above the river; F. H. Knowlton, August 27, 1888.

Yellowstone River, one-half mile below mouth of Elk Creek, top of bluff; F. H. Knowlton, August 27, 1888.

Bluff on Yellowstone River, 1 mile below mouth of Elk Creek; F. H. Knowlton, August 4, 1888.

Cliff on Yellowstone River, short distance above mouth of Hellroaring Creek; F. H. Knowlton, August 10, 1888.

Southeast side of hill above Lost Creek, bed No. 1; F. H. Knowlton, August 8, 1888.

Southeast side of hill above Lost Creek, bed No. 2; F. H. Knowlton, August 8, 1888.

Southeast side of hill above Lost Creek, beds No. 3, 4, 5; F. H. Knowlton, August 8, 1888.

Southern end of Crescent Hill, 300 feet above wagon road; F. H. Knowlton, August 9, 1888.

Southern end of Crescent Hill, "Platanus bed"; F. H. Knowlton, August 9, 1888.

Northeast side of Crescent Hill, opposite small pond in slope, altitude about 7,500 feet; F. H. Knowlton and G. E. Culver, August 2, 1888.

The Thunderer, opposite Soda Butte; F. H. Knowlton and G. E. Culver, August 29, 1888.

East bank of Lamar River, between Cache and Calfee creeks; F. H. Knowlton and G. E. Culver, August 24, 1888.

Hill on road just above Yanceys; F. H. Knowlton, August 6, 1888.

Hill near the Yancey fossil trunks; F. H. Knowlton, August 28, 1888.

Mount Everts, near summit of west end; F. H. Knowlton, July 27, 1888.

Mount Everts, coal opening on side facing the Gardiner River (fragments); F. H. Knowlton, July 26, 1888.

DESCRIPTION OF KNOWN FOSSIL PLANTS FROM THE LARAMIE OF THE YELLOWSTONE NATIONAL PARK.

ASPLENIUM HAGUEI n. sp.

Pl. LXXVII, figs. 1, 2.

Frond thin, delicate, lanceolate in outline, bipinnate, slender, straight; pinnæ alternate, scattered, oblong-lanceolate in shape, cut into few coarse divisions which are either entire or again cut into few obtuse teeth; nervation obscure, consisting of a delicate midvein and few forked branches from it.

This delicate little form is represented by a dozen or more specimens. The longer fragment (fig. 2) is about 4.5 cm. in length and about 1.5 cm. broad. The others are more fragmentary.

Nothing like this has been before reported from the Laramie group. It has some resemblance to *Sphenopteris aequalis* Lx.[1] from the Green River group at Florissant, Colorado, but is much smaller and of decidedly different shape.

It is not certain that it belongs to the genus Asplenium, as no fruit has been observed, but it resembles, at least generically, a number of forms so referred from the Cretaceous of Greenland. For the present it may be retained in this genus.

I have named it in honor of the collector, Mr. Arnold Hague, of the United States Geological Survey.

Habitat: North of Pinyon Peak, on Wolverine Creek, Yellowstone National Park; collected by Arnold Hague, August 10, 1887.

Cret. and Tert. Fl., p. 137, Pl. XXI, figs. 1, 2, 1888.

ONOCLEA MINIMA n. sp

Pl. LXXVII, figs. 11-15.

Fertile frond unknown: sterile frond small, apparently deltoid in outline, deeply pinnatifid into short, rounded, obtuse pinnæ, which are perfectly entire or are cut into few large, coarse teeth; nervation as in the living *O. sensibilis*.

This fine little species is represented by a dozen or more specimens, the best of which are figured. They are all apparently fragments, and consequently it is impossible to make out the real shape of the frond with any certainty. One of the most perfect specimens (fig. 13) is about 3.5 cm. long, and represents the upper portion of a frond or possibly pinnule, if it is a large compound frond. The larger fragment (fig. 15) is 4.5 cm. long and about 1 cm. broad, but it is broken at both ends and there is no means of determining how long it was originally. Fig. 12 at first sight seems to be entirely different from the others, but on comparing it with fig. 14 the only difference observable is that one is cut into a few coarse teeth and the other is entire. The nervation seems to be the same in all and to be identical with that of the living sensitive fern.

Regarding this interesting species, I am somewhat uncertain as to the shape of the frond, and less so as to the genus to which it belongs. Two of the most perfect forms (figs. 11, 13) seem to have come from the upper portion of a frond similar in general shape to the sterile frond of *Onoclea sensibilis*; but, on the other hand, figs. 12 and 14 have much the appearance of being deeply lobed pinnæ, resembling some of the lower ones in *O. sensibilis*. More material will be necessary to settle this point, but in the meantime the species is characteristic enough to be readily distinguishable, and hence is available for geological purposes.

This species was at first thought to be identical with a plant that has been described under the MS. name of *Woodwardia crenata*, which comes from Point of Rocks, Wyoming. This latter is known only from a mere fragment, however, and if additional material could be obtained it might show them to be the same. At present *W. crenata* may be distinguished as being much larger and in having undulate-crenate margins which are minutely serrate. The nervation is practically the same in both.

Onoclea minima has some resemblance to *O. sensibilis fossilis* from the Fort Union group, near the mouth of the Yellowstone. It differs in being hardly one-fifth the size and in having the lobes obtuse and coarsely toothed instead of acute and entire. The nervation is nearly the same in both.

The species under consideration has also the same nervation as *Woodwardia præcedata* from Crescent Hill, but differs essentially in size and shape.

The resemblance to *Woodwardia latiloba* Lx., from the Denver group of Colorado, is still more remote.

Habitat: North of Pinyon Peak, on Wolverine Creek, Yellowstone National Park; collected by Arnold Hague, August 10, 1887. (Field No. 5035.)

ANEMIA SUBCRETACEA (Sap.) Gard. and Ett.

Anemia subcretacea (Sap.) Gard. and Ett.: Monogr. Brit. Eoc. Fl. Vol. I, Pt. II, p. 15, Pl. VIII, Pl. IX, 1880.
Gymnogramma haydenii Lx.: Ann. Rept. U. S. Geol. and Geog. Surv. Terr., p. 295, 1871 (1872); Tert. Fl., p. 59, Pl. V, figs. 1-3, 1878.

The type locality of this species is described as "Divide between the source of Snake River and the southern shore of Yellowstone Lake." It has not since been found inside the Park.

Habitat: As above given.

SEQUOIA LANGSDORFII? (Brgt.) Heer.

Pl. LXXVII, fig. 5.

It is with some hesitation that I refer this fragment to this species. It is small and not well preserved, but the leaves appear to be decurrent and to approach closer in character to this species than to any other with which I am familiar.

Habitat: North of Pinyon Peak, on Wolverine Creek, Yellowstone National Park; collected by Arnold Hague, August 10, 1887.

SEQUOIA REICHENBACHI (Gein.) Heer.

Sequoia reichenbachi (Gein.) Heer: Flor. Foss. Arct., Vol. I, p. 83, Pl. XLIII, figs. 1d, 2b, 5a, 1868.
Abietites dubius Lx. ex p. Lesquereux: Tert. Fl., p. 84, Pl. VI, figs. 20, 21, 21a. Knowlton: Bull. U. S. Geol. Surv. No. 105, p. 46, 1893.

Two small worn fragments are referred to this species. They are obscure, but with little doubt are correctly referred to this form.

Habitat: Mount Everts; about 100 feet above coal mine on west end, below Mammoth Hot Springs; collected by F. H. Knowlton, July 26, 1888.

PHRAGMITES FALCATA n. sp.

Pl. LXXVIII, fig. 5.

Leaves narrowly lanceolate, with a long acuminate apex; nerves rather sparse, about ten in the width of the leaf; intermediate nerves obsolete.

This species rests upon the fragment figured, and, scanty as the material is, differs markedly from the species with which it is associated and to which it is most closely related—that is, *P. alaskana* Heer.

The fragment is 8 cm. in length and 11 mm. in width. It tapers for a distance of 5 cm. to a long, sharp point, thereby differing from *P. alaskana*, which is "obtuse or obtusely mucronate." The primary nerves are 1 mm. apart and reasonably distinct. The secondary or fine nerves can not be made out, owing to the poor state of preservation.

Habitat: Mount Everts, near summit of west end; collected by F. H. Knowlton, July 27, 1888.

GEONOMITES SCHIMPERI Lx.

Geonomites schimperi Lx.: Tert. Fl., p. 116, Pl. X, fig. 4 (1878).
Sabal major? Ung.; Lesquereux: Fifth Ann. Rept. U. S. Geol. and Geog. Surv. Terr., p. 295, 1871 (1872).

This species was collected with *Anemia subcretacea*, and the specimens on which it is based are preserved in the United States National Museum. The species has not been since collected.

Habitat: "Divide between the source of Snake River and the southern shore of Yellowstone Lake."

MYRICA BOLANDERI? Lx.

Pl. LXXVIII, fig. 4.

Myrica bolanderi Lx.: Tert. Fl., p. 133, Pl. XVII, fig. 17 (1878).
Ilex undulata Lx.?[1] Seventh Ann. Rept. U. S. Geol. and Geog. Surv. Terr., 1873 (1874), p. 416.

I refer this single fragment with some hesitation to this species. It differs slightly from the type specimen, which also appears to be the only one thus far mentioned. The one under discussion is about the same size

[1] When this was transferred to Myrica, the specific name *undulata* became preoccupied by *M. undulata* Heer. Schimp., Pal. Veg., Vol. II, p. 516 (1870-1872).

and has the same toothing in the upper portion, differing only in being a little more acute than the type. The basal portion is wanting.

The locality which afforded the original specimen is unknown (cf. Tert. Fl., p. 153), but from the fact that it was sent to Lesquereux with a lot of material from near Florissant, Colorado, it was assumed to belong to the Green River group. It is preserved in the collection of the United States National Museum (No. 1652), and appears to have actually come from the Florissant shale.

Habitat. Mount Everts, near summit, on western end; collected by F. H. Knowlton, July 27, 1888.

QUERCUS ELLISIANA Lx.

Pl. LXXVII, fig. 6.

Quercus ellisiana Lx.: Fifth Ann. Rept. U. S. Geol. and Geog. Surv. Terr. 1871 (1872), p. 297; Tert. Fl., p. 155, Pl. XX, figs. 4, 5, 7, 8, 1878.

A considerable number of specimens that leave no doubt as to the correctness of their determination.

The example figured is only partially preserved and is much larger than is usual in this species. It has, however, the shape and nervation of *Q. ellisiana*, and I refer it with some hesitation to this form.

Habitat: Mount Everts, near the summit of the west end; collected by F. H. Knowlton, July 27, 1888. The figured specimen was collected by George M. Wright, July 7, 1885, on the top of Mount Everts, on the west face.

MALAPOENNA WEEDIANA? Kn.

Malapoenna weediana Kn.: Bull. U. S. Geol. Surv. No. 152, p. 112, 1898.

Litsea weediana Kn.: Bull. U. S. Geol. Surv. No. 105, p. 55, 1893.

Tetranthera sessiliflora Lx. ex. p. Lesquereux: Tert. Fl., p. 217, Pl. XXXV, fig. 9, 1878.

There is a single, much broken fragment that appears to belong to this species. It is too fragmentary to be positive.

Habitat: Top of Mount Everts, west face; collected by George M. Wright, July 7, 1885.

PALIURUS MINIMUS n. sp.

Pl. LXXVII, figs. 7-9.

Leaves thin, membranaceous, nearly circular in outline, very slightly wedge-shaped at base, rounded and obtuse at apex; margin perfectly entire; equally five-nerved from the base, with an occasional branching of the outside nerve, making the leaf appear seven-nerved; midnerve thin, straight, passing to the upper bowler; other nerves of same strength, camptodrome, arching in bows and joining the midvein or midrib; lateral branches few, at an acute angle; finer nervation not preserved.

This fine characteristic species is represented by a number of fairly well preserved examples, the best of which are figured. They are about 2.8 cm. in length and about the same in width. They are nearly circular in shape, being slightly wedge-shaped at the base, but perfectly obtuse at the apex. The nerves are all of about equal strength and divide the space of the blade into approximately equal areas. They occasionally branch, especially the thin central ones.

This species is undoubtedly quite closely allied to several described forms. From *Paliurus colombi* Heer[1] it differs in shape and nervation. It is very much like some of the small leaves of *P. zizyphoides* Lx.[2] from the Laramie of Erie, Colorado, and Black Buttes, Wyoming, but they differ in having the nerves arising from the midrib well above the base of the blade. The species under discussion has precisely the same shape and much the same nervation as *Zizyphus meekii* Lx.[3] but this differs essentially in having a dentate margin.

On the whole it may be characterized by its circular shape, entire margin, and five nerves from the base.

Habitat: North of Pinyon Peak, on Wolverine Creek, Yellowstone National Park; collected by Arnold Hague, August 10, 1887.

PALIURUS ZIZYPHOIDES[?] LX.

PL. LXXVIII, fig. 5.

Paliurus zizyphoides Lx.: Tert. Fl., p. 274, Pl. LI, figs. 1–6, 1878.

This very small leaf is broken at the base and otherwise obscure, but seems to belong to this species.

Habitat: North of Pinyon Peak, on Wolverine Creek; collected by Arnold Hague, August 10, 1887.

Lesquereux, Tert. Fl., p. 273, Pl. L, figs. 13–17.
Op. cit., Pl. LI, fig. 2.
Op. cit., p. 275, Pl. LI, figs. 10–11.

DOMBEYOPSIS PLATANOIDES LX.

PL. LXXVIII, fig. 1.

Dombeyopsis platanoides Lx.: Tert. Fl., p. 254, Pl. XLVII, figs. 1, 2, 1878.

A single specimen.

Habitat: Top of Mount Everts, west face; collected by George M. Wright, July 7, 1885.

ANDROMEDA GRAYANA Heer.

Andromeda grayana Heer: Fl. Foss. Alaska, p. 51, Pl. VIII, fig. 5. Lesquereux: Tert. Fl., p. 231, Pl. XL, fig. 1. Knowlton: Bull. U. S. Geol. Surv. No. 105, p. 56.

Habitat: Mount Everts, near the summit of the west end; collected by F. H. Knowlton, July 27, 1888.

TRAPA? MICROPHYLLA LX.

PL. LXXVII, figs. 3, 4.

Trapa? microphylla Lx.: Tert. Fl., p. 295, Pl. LXI, figs. 16-15a. Ward: Types Laramie Fl., p. 64, Pl. XXVIII, figs. 2-5.

This is undoubtedly the same species as that figured by Lesquereux from Point of Rocks, Wyoming, and by Ward from Burns Ranch, on the lower Yellowstone. It shows more the habit of the specimen figured by Ward, but has the general nervation of all the specimens referred to this species.

In fig. 15a of Tertiary Flora the leaflets are petioled, while in fig. 15 they are clearly similar to Professor Ward's examples.

These curious but well-marked leaves can not possibly belong to the genus Trapa as we now understand it, but as I am at present absolutely unable to suggest any other affinity, I can do nothing but leave their correct determination to be settled by future workers.

Habitat: North of Pinyon Peak, on Wolverine Creek; collected by Arnold Hague, August 10, 1887.

DIOSPYROS STENOSEPALA Heer.

Diospyros stenosepala Heer. Lesquereux; Fifth Ann. Rept. U. S. Geol. and Geog. Surv. Terr., p. 296, 1871 (1872).

Habitat: "Divide between the source of Snake River and the southern shore of Yellowstone Lake."

FRAXINUS DENTICULATA Heer.

Pl. LXXVIII, fig. 6.

Fraxinus denticulata Heer; Fl. Foss. Arct., Vol. I, p. 118, Pl. XVI, fig. 4; Pl. XLVII, fig. 2. Lesquereux; Tert. Fl., p. 228, Pl. XL, figs. 1, 2.

Several well-preserved specimens that are referred with certainty. Besides these there are several other well-preserved examples, of which the one figured is perhaps the best, that are somewhat larger than the types, but still appear to belong with them. The nervation is obscure, but the shape and toothed margin are quite similar.

Habitat: Mount Everts, near summit of the west end; collected by F. H. Knowlton, July 27, 1888.

VIBURNUM ROTUNDIFOLIUM Lx.

Pl. LXXVII, fig. 10; Pl. LXXVIII, figs. 2, 8, 9.

Viburnum rotundifolium Lx.; Tert. Fl., p. 225, Pl. XXXVII, fig. 12; Pl. XXXVIII, fig. 10; Pl. LXI, fig. 22.

There is considerable difference in size among the specimens, but they seem to belong together, and to approach quite closely to Lesquereux's species. The small leaf shown in fig. 9, for instance, is certainly the same as the plant figured by Lesquereux (loc. cit., Pl. LXI, fig. 22), while fig. 8 is like fig. 10, Pl. XXXVIII (loc. cit.).

Habitat: North of Pinyon Peak, on Wolverine Creek; collected by Arnold Hague, August 10, 1887.

Table showing geological distribution of Laramie plants.

	Yellowstone National Park.				Outside.				
	Mount Everts.	Wolverine Creek.	Divide between Snake River and S. part of the Lake.	Locality of Watonga(?) Creek, Lewis and New Mexico.	Jaw near Laramie.	Burlington.	Denver.	Fort Union(?) river beds.	Green River group.
Asplenium haguei n. sp.		×							
Onoclea minima n. sp.		×							
Anemia subcretacea (Sap.) Gard. and Ett.			×	×	×		× ?		
Sequoia langsdorfi (Brgt.) Heer		×						×	
Sequoia reichenbachi (Gein.) Heer	×			×		×			
Phragmites fabata n. sp.	×								
Gymnomites schimperi Lx.		×							
Myrica bolanderi ? Lx.	×								×
Quercus ellisiana Lx.	×				×				
Malapoenna weediana ? Kn.	×					×			
Paliurus minimus n. sp.		×							
Paliurus zizyphoides ? Lx.		×		×					
Bumelyopsis platanoides Lx.	×				×				
Trapa ? microphylla Lx.		×						×	
Andromeda grayana Lx.	×				×	×			
Diospyros stenosepala Heer a		×							
Fraxinus denticulata Heer	×				×	×			
Viburnum rotundifolium Lx.		×		×					
	8	7	3	5	5	4	1?	2	1

a Very doubtful

DISCUSSION OF LARAMIE FLORA.

It will be observed that there are only three localities within the Yellowstone National Park that have afforded Laramie plants, viz: Near the summit of Mount Everts, the valley of Wolverine Creek, and the more or less doubtful locality known as "divide between the source of Snake River and the southern part of the Yellowstone Lake." It has not been possible to relocate the latter place, but as it is in a region in which Laramie strata are known to occur, and several of the species represented have since been found in Laramie strata outside, it is assumed to be correct.

This flora embraces only 18 species, of which number 8 are confined to the Mount Everts locality, 7 to Wolverine Creek, and 3 to the above-mentioned doubtful locality.

Of the 8 species found at Mount Everts, 1 (*Phragmites falcata*) is described as new, and 2 species (*Quercus ellisiana* and *Dombeyopsis platanoides*) have never before been found outside of the so-called Bozeman Laramie. The 2 species regarded as doubtful (*Myrica halawleri* and *Malapœnna weediana*) depend on a single fragment each and are obviously of no value in determining the age. They are found normally in much higher horizons. Of the 3 remaining species, *Sequoia reichenbachi* has been found in the Livingston beds, but is also found in the true Laramie, and abundantly in still older strata. *Andromeda grayana* and *Fraxinus denticulata* have been found in both Laramie and Livingston beds in the Bozeman area.

The evidence of the fossil plants confirms that derived from the study of the stratigraphy, namely, that the beds near the summit of Mount Everts are of Laramie age.

Of the 7 species from Wolverine Creek, 3—*Asplenium knowltoni, Onoclea minima*, and *Palinrus minimus*—are described as new. The first of these does not appear to have any very close relatives in North America, but apparently finds its nearest analogue in certain species from the Cretaceous of Greenland. *Onoclea minima*, on the other hand, is very close indeed to a fern from Point of Rocks, Wyoming, that has been described under the manuscript name of *Woodwardia crenata*. The Wyoming plant depends on two or three small fragments, which, as pointed out under the diagnosis of *Onoclea*, are hardly sufficient to properly characterize it. It is quite possible that when new material shall be obtained these two plants will be found identical. *Palinrus minimus* is perhaps nearest to *P. zizyphoides* from Black Buttes, Wyoming, and Erie, Colorado. The 4 remaining species are distributed as follows: *Sequoia langsdorfii* is represented by 1 small branchlet, and the identification is probably correct, as it is an easily recognized species. It has a wide geological and geographical distribution, being especially abundant in the lower Fort Union beds. *Palinrus minimus* is doubtfully identified in this material. As stated above, it is a true Laramie species. *Viburnum rotundifolium* is also a Laramie species. It has never before been found outside of the type locality, which is Point of

Rocks and Black Buttes, Wyoming. The most interesting species is *Ficus? microphylla*. It is represented by several perfectly characteristic specimens. This species was first described from Point of Rocks, Wyoming, and was found later by Professor Ward in lowest Fort Union beds, near the mouth of the Yellowstone River. The Wolverine Creek specimens approach closest to Professor Ward's specimens. Professor Ward is of the opinion that these lower beds represent the Laramie, since the plants in them differ from those in the undoubted Fort Union beds.

The three species from the divide between Snake River and the southern part of Yellowstone Lake are of little value in determining the age. *Ceanomites schimperi* has never been found in any other locality, and *Diospyros stenosepala* is very doubtful indeed. It has not since been collected, and the specimen on which Lesquereux based his determination can not now be found. The only remaining species, *Asplenium subcretacea* or *Gymnogramma haydenii* of Lesquereux, has a wide distribution, having been found in the Laramie, Denver, and Eocene.

DESCRIPTION OF FOSSIL PLANTS FROM THE TERTIARY OF THE YELLOWSTONE NATIONAL PARK.

PLANTS, EXCLUSIVE OF FOSSIL WOOD.

FILICES.

WOODWARDIA PREAREOLATA n. sp.

Pl. LXXIX, fig. 1.

Frond pinnate; pinnae alternate, lanceolate, with slightly undulate margins, connate at their bases, forming a broad wing on the rachis; nervation strongly reticulated, consisting of one or two rows of long lacunae next to the main rachis and along the secondary rachis, and the remainder forming large polygonal, slightly elongated, meshes.

Unfortunately the specimen figured represents the only example found. It is far from perfect, being only a segment from the middle of a frond, and consequently no idea can be gained of the outline of the whole frond. The segment of the main rachis is 8 cm. long. The pinnae are regularly alternate—that is, are the same distance apart. They are at least 6 cm. in length and 2 cm. in width. The full length and form of the apex could not be determined. The nervation is well shown in the figure.

This species is undoubtedly very closely related to the living *Woodwardia areolata* (L.) Moore—so closely, in fact, that it is hardly possible to separate them satisfactorily. The pinnae are alternate, of the same shape, and have identical nervation in both. The only difference is that the margins of the pinnae are entire in the fossil and more or less serrate in the living species. It is possible that more material of the fossil form would show differences in this respect and bring them absolutely together.

This new fossil species much resembles *Onoclea sensibilis fossilis* Newberry, from the Fort Union group, but it differs in having strictly alternate pinnae that are as far apart as it is possible to be. The pinnae are also without lobes of any kind, being only slightly undulated. The nervation differs slightly in producing more elongated areolae in *O. sensibilis fossilis*.

In nervation *W. proeoalata* resembles *W. latiloba* Lx.,[1] from the Denver group, but differs markedly in having the pinnae unlobed.

Habitat: Northeast side of Crescent Hill, opposite small pond; collected by F. H. Knowlton and G. E. Culver, August 2, 1888.

ASPLENIUM BOOMSTI n. sp.

Pl. LXXIX, figs. 2, 3; Pl. LXXX, figs. 9, 10.

Frond large, at least twice pinnate; main rachis thick, slightly zigzag; pinnae alternate, remote, standing at an angle of 30° to 45°, long-lanceolate, tapering to a rather slender apex, rachis strong, often abruptly curved upward, cut into numerous alternate, oblong, obtusely-acuminate pinnules with upward-turning points; nervation of pinnules simple, consisting of a slender midnerve and about 9 pairs of unforked, close, parallel branches, which are slightly arched forward in passing to the borders; sori oblong, nearer the margin than the midnerve.

This fine species is represented by a large number of well-preserved specimens. It appears to have been a very large fern, possibly several times pinnate, but none of the specimens show the larger connections. The largest rachis with pinnae attached is 4 mm. thick, but on the same stone, and

Tert. Fl., p. 54, Pl. III, figs. 1-1c.

evidently of the same species, are stems or rachises fully 8 m m. thick. There is some evidence to indicate that they were combined into a very large frond, but it is not conclusive. The longest example is about 20 cm. long and spreads about 9 cm. The longest pinna (Pl. LXXX, fig. 9) is preserved for 9 cm. and still lacks the terminal portion. The pinnæ vary in width from 10 to 24 mm., depending upon the portion taken.

The pinnæ are cut into oblong acuminate pinnules, the sinus sometimes extending to within one-third of their length of the base, but usually to about half the length. Pinnules with a slender midnerve and 7 to 10, usually 9, pairs of close, unforked nerves. The lower nerves of adjoining pinnules unite at a low angle and pass upward and end in the sinus.

Fruit dots were observed only on one small fragment (Pl. LXXIX, fig. 2). They are obscure, but as nearly as can be made out they are oblong and near the margin of the pinnules. Unfortunately none of the larger specimens are fruiting, but apparently they all belong to the same species.

I do not recall any fossil species to which this seems to be allied. A number that have been described resemble it, but none closely enough to constitute specific similarity.

I have named this species in honor of Prof. Joseph P. Iddings, of the University of Chicago, who pointed out the locality which afforded the best specimens.

Habitat: Yellowstone River, above mouth of Hellroaring Creek (figs. 5, 10), bank of Yellowstone, one-half mile below mouth of Elk Creek, base of bluff (figs. 2, 9); collected by F. H. Knowlton from Fossil Forest Ridge, bed No. 6, "Platanus bed," August, 1888. One specimen collected by Lester F. Ward and F. H. Knowlton, August, 1887.

ASPLENIUM MAGNUM n. sp.

Pl. LXXIX, figs. 5-8, 8a.

Frond simple, pinnatifid, sometimes nearly pinnate below, long lanceolate in outline, from a regular obtusely wedge-shaped base, and extending into a long slender apex; cut into numerous, mainly alternate, ovate, sharp-pointed lobes, those at the base being sometimes cut nearly to the rachis, those above less and less until the apex is nearly entire; nervation of the lobes or pinnules consisting of a strong midnerve passing to the tip, and 6 or 8 pairs of alternate once-forked lateral nerves; fruit dots not seen.

This large and striking species is the most abundant fern found in the Park. It is represented in the collection by fully 40 specimens, all from one locality. The largest example (fig. 7) is 17.5 cm. long and 2.5 cm. broad, and still lacks the terminal portion. It has a stipe 8 mm. long and 2 mm. thick. Fig. 5 is 16 cm. long and 23 mm. broad, and lacks both base and apex. Some of the fronds must have been fully 25 cm. long.

This species is well characterized. It has a thick grooved rachis (1 mm.) and a short thick stipe. The lobes or pinnules are irregularly ovate, separated usually to the middle by a sharp sinus, and having a sharp upward-pointing apex. The nervation consists of a strong midnerve ending in the apex and about 7 pairs of alternate forked branches. As in the former species, the lower nerves of adjacent lobes unite and pass upward to the sinus. Occasionally there may be an unforked nerve, but it is the exception.

This species is associated in the same beds with *A. iddingsi* and much resembles it, differing, however, in having apparently simple fronds that are uniformly larger than the pinnæ of the former species, and in having the nervation of the lobes always forking (see fig. 8 and 8a). It differs further in having a short stipe and in having the upper portion nearly or quite entire.

The correctness of this generic reference is of course a matter of more or less doubt, as no fruiting specimens have been found, but the fern appears to be allied generically at least to *A. iddingsi*. It is certainly a well-marked species for geological purposes.

Habitat: Yellowstone River, one-half mile below mouth of Elk Creek, Yellowstone National Park, at base of bluff; collected by F. H. Knowlton, August, 1888.

ASPLENIUM EROSUM? (LX.) KN.

PL. LXXX, fig. 6.

Asplenium erosum (LX.) KN.: Bull. U. S. Geol. Surv. No. 152, p. 45, 1898.
Pteris erosa LX.: Tert. Fl., p. 53, Pl. IV, fig. 8.

This appears to be the same as described by Lesquereux, but is obscure and difficult to make out. None of the specimens are complete, and all have the nervation very poorly preserved. The margin seems more erose than the type, and the nerves may not fork. It is possible that it is a new species, but until better material can be obtained I have preferred to retain the specimens under this species.

Habitat: Yellowstone River, one-half mile below mouth of Elk Creek; collected by F. H. Knowlton, August 13, 1888 (fig. 6). Yellowstone River, wall of canyon above mouth of Hellroaring Creek, collected by W. H. Weed, October 13, 1887.

ASPLENIUM REMOTIDENS n. sp.

PL. LXXX, fig. 7.

Pinnæ large, coriaceous, broadly lanceolate, taper-pointed, obtuse and unequal-sided at base; margin with few remote sharp teeth; midvein strong; lateral veins at an angle of about 45°, simple or forking once some distance above the midvein; sori not seen.

The very perfect example figured is the only specimen obtained. It is 11 cm. in length and 2.5 cm. broad. It is broadly lanceolate with a slender tapering apex and obtuse unequal-sided base. The nervation is very obscure. It is probable that all of the lateral veins fork, but it was not possible to make this out, and the figure shows many as unforked. The ones that are made out to have the fork show it some distance above their base.

This species is very closely allied to, if not indeed identical with, *Asplenium crosum* (Lx.) Kn.,[1] from the Denver formation of Colorado. It has exactly the same shape, but differs in having few remote teeth, and in the branching of the veins. In *A. crosum* the veins fork at the base and occasionally above the middle. In any case the species are very close together and may be combined at any time if future material from the Yellowstone National Park shows variation in the characters now relied upon for their separation.

Habitat: Yellowstone River, one-half mile below mouth of Elk Creek, at base of bluff; collected by F. H. Knowlton, August, 1888.

DRYOPTERIS WEEDII n. sp.

PL. LXXX, fig. 8; PL. LXXXI, fig. 2.

Frond twice pinnate; pinnæ probably lanceolate in outline; pinnules

[1] Under *Pteris crosa* Lx., Tert. Fl., p. 53, Pl. IV, fig. 8; Cret. and Tert. Fl., p. 124, Pl. XIX, fig. 1.

opposite or subopposite, nearly at right angles to the rachis, long-lanceolate, rather abruptly acuminate, cut to within a very short distance of the rachis; nervation simple, consisting of a strong straight midvein and numerous (about 20) pairs of opposite, parallel, unbranched, lateral nerves; fruit dots small, round, on the backs of the nerves midway between the midvein and the margin.

This beautiful species is represented by several specimens, the best of which is shown in fig. 8. The pinnules are opposite or subopposite. They are long, slender, and pointing upward. The nervation is very regular, consisting of the strong midvein and 18 to 20 or more pairs of opposite parallel veins. The fruit dots are distinct, though small, and borne on the veins midway between the midvein and margin.

This species is closely allied to *Lastrea goldiana* (Lx.) Kn.[*] from the Denver beds of Golden, Colorado. It does not so closely resemble the figure given by Lesquereux as it does certain forms that have been referred to it in my forthcoming monograph of the Laramie and allied formations. The type of the species is described by Lesquereux as having 5 to 7 pairs of nerves, while the forms that I have referred to it have 10 pairs, with no other apparent difference. *Dryopteris weedii*, as stated, has 18 to 20 or more pairs. The pinnules are from 10 to 16 mm. long and about 5 mm. broad, whereas those of *L. goldiana* are only 7 to 9 mm. long and 3 or 4 mm. broad.

From this it is clear that these two species are quite closely related, and possibly a larger series of specimens might show them to be identical, but for the present it is best to regard them as different.

I have named this species in honor of Mr. Walter Harvey Weed, by whom the first specimens were collected.

Habitat: Yellowstone River, breccia in wall of canyon above mouth of Hellroaring Creek (Pl. LXXXI, fig. 2); collected by Walter Harvey Weed, October 13, 1887. Cliff on Yellowstone River (left hand), short distance above mouth of Hellroaring Creek (Pl. LXXX, fig. 8); collected by F. H. Knowlton, August 10, 1888.

---

[*] This was first called *Lepidoid goldianus* by Lesquereux, Seventh Ann. Rept., 1873 p. 393, but was later changed to *Lastrea (Goniopteris) goldiana* (Cr. Tert. Fl., p. 56, Pl. IV, fig. 13).

DRYOPTERIS XANTHOLITHENSIS n. sp.

Pl. LXXXI, fig. 4.

Frond pinnate?; pinnae lanceolate; pinnules opposite, lanceolate-deltoid, obtuse, cut to within one-third of their length of the base, much arched upward at the point; nervation simple, consisting of well-marked midvein and 9 or 10 pairs of opposite, parallel, unbranched lateral veins; sori large, round, on the backs of the veins at about one-third of their length from the midvein.

Of this well-marked species the single specimen figured was the only one found. It is not preserved entire, the fragment being about 5 cm. in length. There is therefore no means of knowing whether or not it was simple or compound. The portions of the pinnae preserved are of the same width throughout, showing that they probably came from the middle portion. The pinnules are opposite and arise at an angle of 30° or 40° from the rachis. They are lanceolate-deltoid in shape, and about 12 mm. long and 5 mm. broad, being much arched upward at the extremity. The fruit dots are large, round, and placed on the backs of the veins near the midvein.

This species is allied to *Dryopteris wordii*, from which it clearly differs in having much shorter, arching pinnules, only 9 or 10 pairs of nerves, and larger fruit dots which are nearer the midvein. The nervation is the same in character, but differs, as stated, in number of pairs of veins.

From *Lastrea goldiana* this species differs in much the same manner. It has more arching pinnules, and is quite different in general appearance. The number of pairs of nerves is, however, about the same; all of which goes to show that these three species are closely related.

Habitat: Fossil Forest Ridge, Yellowstone National Park, bed No. 6, "Platanus bed;" collected by Lester F. Ward and F. H. Knowlton, August 19, 1887.

DAVALLIA? MONTANA n. sp.

Pl. LXXIX, fig. 4.

Frond thin, twice pinnate, possibly more compounded; rachis strong; pinnae alternate, lanceolate, ending in a sharp, hair-like point; cut into 5 or 6 lobes or pinnules, the sinuses toward the base going nearly to the secondary rachis, more entire near the apex; lower pinnules or divisions 2 or 3 toothed, all, but especially the terminal pinnules, ending in long, slender, outward-pointing teeth; nervation simple, consisting of a strong secondary rachis with rather delicate nerves pointing to the pinnules; the nerves near the base two or three times branching, the branches entering the teeth; nerves near the extremity unbranched.

The small fragment figured represents all that was found of this species. It is only about 30 mm. long and 25 mm. broad, the pinnæ being 17 mm. in length and approximately 10 mm. in width.

Notwithstanding the smallness of the fragment, there is enough to show that it differs markedly from any other form in the Park flora. I am quite at loss, however, to indicate its generic affinities. I have placed it under Devallia tentatively, and can only hope that subsequent material will serve to fix more satisfactorily its position.

Habitat: Fossil Forest Ridge, Yellowstone National Park, bed No. 3, "Magnolia bed," collected by Lester F. Ward and F. H. Knowlton, August 16-19, 1887.

LYGODIUM KAULFUSII Heer.

Pl. LXXX, figs. 1–3.

Lygodium neuropteroides Lx.: Tert. Fl., p. 64, Pl. V, figs. 4–7; Pl. VI, fig. 1.

According to Gardner,[1] the *Lygodium neuropteroides* of Lesquereux is absolutely identical with *L. kaulfusii* of Heer. Lesquereux was shown specimens of the true *L. kaulfusii* from the British Eocene and pronounced them "positively identical" with his species from the Green River group and later formations. A glance at Gardner's[2] figures shows that it is impossible to separate the American specimens.

This species was found at two localities in the Yellowstone National Park, namely, on the Yellowstone River below the mouth of Elk Creek, and on the north bank of the Lamar River between Cache and Calfee creeks. The specimens from below Elk Creek are in hard, rather coarse-grained rock at the base of the section. They are very large, having lobes

---

Brit. Eoc. Fl., Pt. I, Ferns, p. 47.   Op. cit., Pl. VII, figs. 1, 3, 8; Pl. X, fig. 11.

8 cm. long and 2 cm. broad, and very much resemble a figure of this species given by Newberry,[1] from the Pacific coast.

The specimens from the Lamar River are much slenderer, being 7 cm. long and less than 1 cm. broad. Some of them, as fig. 2, are very small and delicate. In nervation the specimens from both localities agree perfectly, as they do with European specimens.

Habitat: Yellowstone River, one-half mile below mouth of Elk Creek, at base of section; collected by F. H. Knowlton, August 13, 1888. North bank of Lamar River, between Cache and Calfee creeks; collected by F. H. Knowlton, August 24, 1888.

OSMUNDA AFFINIS LX.

Pl. LXXX, figs. 4, 5.

Osmunda affinis Lx.; Tert. Fl., p. 59, Pl. IV, fig. 1.

The Park collection contains specimens of several detached pinnules that it seems necessary to refer to this species. They are about the same size as Lesquereux's type specimen, but are better, in that they show the bases of the pinnules.

In the collection of Denver group plants recently worked up there are a number of specimens of this species, some of which are very fine. One in particular, which has been figured for the forthcoming monograph of the Laramie and allied formations, is very large and perfect. It has a long zigzag rachis with numerous sessile pinnules alternately attached. They have a distinctly heart-shaped base, a slightly undulate margin, and a tapering but obtuse apex. In all these particulars, as well as in nervation, the Park specimens agree. The latter are a little shorter than the Denver specimens, and one is a trifle broader, but the differences are unessential. There is no knowing the part of the frond from which they came, and this may readily account for discrepancies.

Habitat: Southeast side of hill north of Lost Creek, Yellowstone National Park, bed No. 4, about 6,550 feet altitude; collected by F. H. Knowlton, August 5, 1888.

Plates. Pl. LXII, fig. 4. [Ined.]

EQUISETACEÆ.

EQUISETUM HAGUEI n. sp.

Pl. LXXXI, figs. 3, 4.

Stem simple, striate, articulate; articulations rather long; sheaths short; teeth long, appressed, sharp-pointed.

This species is represented by numerous fragments, many of which, however, show the sheaths. The stem is from 4 to 6 mm. broad and the articulations are 5 to 6 cm. in length. It is plainly striate, with usually 9 ribs. The sheaths, which are darker in color than the stem, are 6 or 7 mm. in length and are provided with closely appressed, very sharp-pointed teeth, about 3 mm. long.

If there are 8 or 9 ribs now visible in the flattened stems, it seems safe to assume, inasmuch as they were cylindrical, that they have 16 or 18 ribs, and an equal number of teeth.

It was at first supposed that these specimens could be referred to *E. limosum* L. [see following species],[1] as identified by Lesquereux from material collected by Hayden from basaltic rocks near the Yellowstone Lake; but an examination of the type specimen preserved in the United States National Museum (No. 41) shows that they can not be the same. Lesquereux's specimen has only 4 or 5 ribs visible, making, as he says, about 10 for the entire diameter, while this has 16 to 18, and possibly as many as 20. The segments of the stem are only about 4 cm. in length in Lesquereux's specimen and 6 or 7 cm. in the one under discussion. The sheaths are also longer and the teeth sharper in *E. haguei*.

Among living species this seems to approach closely to *E. limosum* L.; more closely, in fact, than does the specimen referred to *E. limosum* by Lesquereux.

I have named this species in honor of Mr. Arnold Hague, who pointed out the locality where it was found.

Habitat: Southeastern end of hill north of Lost Creek, Yellowstone National Park, bed No. 5; collected by F. H. Knowlton, 1888.

[1] Fifth Ann. Rept. U. S. Geol. and Geog. Surv. Terr., 1871-1872, p. 386; Tert. Fl., p. 69, Pl. VI, fig. 5.

EQUISETUM LESQUEREUXII KN.

Equisetum lesquereuxii Kn.: Bull. U. S. Geol. Surv. No. 152, p. 94, 1898.
Equisetum limosum Linn. Lesquereux: Fifth Ann. Rept. U. S. Geol. and Geog. Surv.
Terr., 1871 (1872), p. 299; Tert. Fl., p. 69, Pl. VI, fig. 5.

As already stated under *E. laqua i*, the type specimen of this species is in the United States National Museum (No. 44). The figure given in Tert. Fl. (Pl. VI, fig. 5) is much more perfect than the specimen proves to be. The figure shows 7 ribs and the same number of teeth, which would make at least 14 ribs for the whole stem. The specimen shows only 4 or 5 ribs, and the sheaths and teeth are very obscure.

As it seems very unlikely that it should belong to the living species, I have ventured to change it, and have named it in honor of Professor Lesquereux.

Habitat: "Near Yellowstone Lake, among basaltic rocks."

EQUISETUM CANALICULATUM n. sp.

Pl. LXXXI, figs. 6, 7.

Stem large, about 50-ribbed; articulate; articulation long; sheath obscure, but apparently short; teeth numerous, short-appressed, sharp-pointed.

This species is represented by the two fragments figured and a number of other doubtful ones, which are hardly enough to properly characterize the species; but they seem to differ from all described species likely to occur in this region, and I have ventured to give them a new name. More perfect material may bring out the relationship.

The longest stem (fig. 6) is about 6 cm. in length, and the broadest on that piece of material is 13 mm. The other specimen (fig. 7) is 5 cm. long and 21 mm. broad. The ribs are distinct, yet not specially strong. They number, as nearly as can be made out, about 25 on a side, or approximately 50 for the entire diameter. The length of the segments can not be made out. The sheath is also obscure. It may be that fig. 6 represents a single sheath; if so, it is long, but the other specimen gives slight evidence of having a short sheath. The teeth are short and appressed and end in slender points. As near as can be made out, there are about 25 teeth in view, or something like 50 for the whole stem.

The two stems shown in the upper part of fig. 6 show a distinct line of tubercles about the slight constriction. They probably represent the lower portions of the stems.

Habitat: Yancey Fossil Forest, beds near the upright stumps (fig. 6); collected by F. H. Knowlton, August 28, 1888. End of Specimen Ridge, opposite Junction Butte, near large upright stumps; collected by Lester F. Ward and E. C. Alderson, August 25, 1887. Yellowstone River, one-half mile below Elk Creek, at base of bluff; collected by F. H. Knowlton, August 27, 1888. Cliff west of Fossil Forest Ridge; collected by Ward and Knowlton, August 15, 1887.

EQUISETUM DECIDUUM n. sp.

PL. LXXXI, fig. 5.

Stems large, many-ribbed, articulate, sheathed; sheaths short, without teeth.

This form is represented by several specimens, all very fragmentary and obscure. It has the stem 15 mm. in diameter, and the sheath 14 mm. in length. The diaphragm is clearly shown in 2 specimens, and appears to have been thick. The sheath is close and without teeth, which probably indicates relationship of this species with living species, such as *E. hiemale*, *E. robustum*, etc., having deciduous teeth.

Habitat. Yellowstone River, one-half mile below the mouth of Elk Creek, base of bluff (fig. 5); collected by F. H. Knowlton, August 27, 1888. Fossil Forest Ridge, bed No. 6, "Platanus bed;" collected by Ward and Knowlton, August 19, 1887.

CONIFERÆ.

PINUS GRACILISTROBUS n. sp.

PL. LXXX, fig. 12.

Cone lanceolate, about 12 mm. in diameter and about 45 mm. long (neither base nor apex preserved); scales in 7 or 8 rows, regularly rhomboidal in shape, about 6 mm. in transverse and about 5 mm. in vertical dimension; scales umbonate, with usually 3 slight projections on the lower angle.

The specimen figured is the only one found of this species. At first sight it seems hardly possible to have so long and narrow a cone with so large scales, but this cone is preserved entire—that is, it has been pressed flat, and by turning it around the entire series of scales may be made out. It is now pressed into an elliptical shape, with a long diameter of about 12 mm. and a short diameter of about 5 mm. Its length, as already stated is approximately 45 mm.

I have not been able to find any fossil species with which this can be compared. There are a number having scales of much the same shape, but none with the same sized cone.

Habitat: Fossil Forest Ridge, Yellowstone National Park, bed No. 7, "Castanea bed;" collected by Lester F. Ward and F. H. Knowlton, August 16–20, 1887.

PINUS PREMURRAYANA n. sp.

PL. LXXXII, fig. 5.

Cone narrowly ovate-conical, rounded at base and gradually narrowed above to a very obtuse and rounded apex; scales thick, regularly rhomboidal, transversely wrinkled, each provided with a rounded blunt umbo, or possibly with a short, stout spine.

This species is represented by the single specimen figured, and is the most perfectly preserved cone I have ever seen, being preserved entire, with little or no distortion. It is about 8 cm. in length. It is broadest at base, where it is about 2.5 cm. in diameter, from which point it tapers gradually to the apex, where it is about 1 cm. in diameter. The scales are very tightly closed, showing that with little doubt the cone was serotinous. They are quite regularly rhomboidal, being about 10 mm. long and 6 mm. high, and appear to have been transversely wrinkled. The top is thick, raised, and was provided, in all probability, with a short, stout spine.

In seeking the probable affinities of this cone, a number of interesting problems are presented, first of which is the state of maturity. It is, of course, a well-known fact that all cones are tightly closed after fertilization and until the seeds are matured. In the majority of cases the scales open for the discharge of the ripe seeds, yet in a number of species they remain closed, or practically so, for many years. The seeds of these serotinous cones may retain their vitality for years—a provision for the continuance of the species.

Whether the cone under consideration is immature, and has the scales closed on this account, or is a strictly serotinous form, is a difficult matter to decide. On the whole it seems probable that it was nearly or quite mature, and should be placed among those with normally closed scales.

The next question is, What is the age of this cone?—that is, is it a cone of a recent species or does it represent an extinct form? The phenomena are so active in the Park at the present time that it is perhaps possible for a cone of this kind to be replaced with silica within a comparatively short space of time. It however came from a part of the Park where the hot-springs phenomena have ceased for a long period, and this lends color to the idea that it is not of very recent origin. The probability is, therefore, that it represents an extinct rather than a living species.

This cone clearly belongs to the pitch pines and not to the soft or white pines, and in determining its affinities this latter group must be excluded. At the present time there are 3 species of pines growing in the Yellowstone National Park, as follows: *Pinus scopulorum, P. flexilis,* and *P. contorta murrayana.* Of these, *P. flexilis* belongs to the white pines and the others to the so-called pitch pines, and of these the last, or *P. contorta murrayana,* is by far the most abundant.

I have shown this cone to a number of botanists familiar with the present flora, and there seems to be much diversity of opinion as to its probable relationship. Mr. F. V. Coville, botanist of the Department of Agriculture, inclines to regard it as allied to an immature cone of *P. scopulorum,* but a careful comparison fails to sustain this view. Mr. George B. Sudworth, dendrologist of the Department of Agriculture, regards it as most closely allied to *P. contorta murrayana,* the lodge-pole pine, and I have so considered it. It is of approximately the same shape as mature cones of this species, but is longer and rather narrower. It is not improbable, as suggested by Mr. Sudworth, that it represents a form which was the immediate ancestor of *P. contorta murrayana,* and I have given it the tentative name of *premurrayana.*

Habitat: East of the Yellowstone Lake, Yellowstone National Park. Collected by members of the Yellowstone National Park division of the United States Geological Survey.

PINUS sp.

Cone lanceolate?, about 16 mm. in diameter, length of part preserved 18 mm.; scales 5 rows in part preserved, probably about 10 or 12 in whole cone, approximately square (or, better, rhomboidal), 5 mm. in each direction; each scale marked by a distinct mubo, and with a prominent ridge along the lower part.

The specimen described is also the only one observed. It is possible that it may be the same as *P. gracilistrobus*, as it comes from the same beds, but it is nearly twice the size, and differs slightly in the character of the scales and their markings.

Neither the base nor the apex is preserved, and it is therefore impossible to know the length, but there is a slight indication that the part preserved is near the upper end, as it is slightly narrowed. This may, however, be due to the poor state of preservation.

As stated under *P. gracilistrobus*, it is hardly worth while to attempt working out affinities with such imperfect material.

Habitat: Fossil Forest Ridge, Yellowstone National Park, bed No. 7, "Castanea bed;" collected by Lester F. Ward and F. H. Knowlton, August 16–20, 1887.

PINUS MACROLEPIS n. sp.

PL. LXXX, fig. 11.

Scales thick, spatulate, rounded above, slender below, with a raised margin or rim.

The mere fragment figured represents all that was found of this species. It consists of portions of 9 scales, arranged in 4 spiral rows. They are broadly spatulate, being rounded above and narrow below. The largest one is 13 mm. in length, 6 mm. broad in the upper portion, and about 3 mm. in the lower portion. The scales were thick and have a strong raised rim.

There is every evidence that this was a large cone, but it is so fragmentary that nothing can be made out but the few scales. It is useless to attempt to work out affinities, except that it was probably a white pine.

Habitat: Cliff west of Fossil Forest Ridge, Yellowstone National Park; collected by Lester F. Ward and F. H. Knowlton, August 15, 1887.

PINUS WARDII n. sp.

Leaves linear, long, apparently in twos, ribbed, not terete.

There are a considerable number of fragmentary specimens that seem in all probability to belong to this genus. They are slender, needle-like leaves, about 1 mm. broad and at least 8 cm. in length. They appear to have been ribbed, or at least not terete in cross section. In no case has the point or base of the leaves been observed, but from the fact that two leaves seem to be found side by side, it seems quite probable that they were arranged in twos, as in the living *Pinus edulis* Engel., *P. contorta* Dougl., etc.

These leaves have a more or less close resemblance to certain described species, but they are too indistinct and poorly characterized to make any comparison valuable.

Habitat: Fossil Forest Ridge, Yellowstone National Park, bed No. 4, "Aralia bed;" altitude about 8,475 feet; collected by Lester F. Ward and F. H. Knowlton, August 20, 1887.

Pinus iddingsi n. sp.

Pl. LXXXII, figs. 8, 9.

Leaves linear, very long, teretish, ribbed, in bundles of three, sheath obscure, but apparently short.

The collection contains a number of needle-shaped leaves that without doubt belong to Pinus. They are about 2 mm. broad and at least 13 cm. long, but none are preserved entire. They appear to be associated in threes, and in one case (see fig. 8) the upper portion of a sheath is preserved with three leaves arising out of it. As near as can be made out, the leaves are round on one side and flat and channeled on the other.

I have named this species in honor of Prof. Joseph P. Iddings, the collector.

Habitat: Andesitic breccia near gulch northwest of peak west of Dunraven; collected by Joseph P. Iddings, September 12, 1883.

Taxites olriki Heer.

Pl. LXXXII, figs. 1, 4, 5.

Several specimens of this fine species were found. They agree closely with the figures given by Heer[1] of specimens from Atanekerdluk, Greenland.

Habitat: Walls of the canyon of Yellowstone River above mouth of Hellroaring Creek, Yellowstone National Park; collected by Walter Harvey Weed, October 13, 1887. (Field No., 2961.)

Fl. Foss. Arct., Vol. I, p. 95, Pl. I, figs. 21-24.

SEQUOIA COUTTSIÆ Heer.

The material which I incline to refer to this species was found at only one locality within the Park, namely, the northeast side of Crescent Hill. It was abundant and fairly well preserved. It consists of masses of slender branches with short acute appressed leaves, in some cases with recurving or at least spreading tips. In a number of cases the male aments were preserved. They are on short, slender branches covered with short scales. The aments are made up of few small, irregular scales.

There is undoubtedly much confusion in regard to this species. According to Gardner,[1] much of the material referred to by Heer and others, from Greenland especially, should be relegated to another species, which he proposed to call S. rhampeci. Gardner is also of the opinion that portions of the foliage have by various authors been separated as Glyptostrobus ungeri. These, as he points out, are usually associated with Sequoia cones, and are "never accompanied by any trace of the persistent and very distinct cones of Glyptostrobus." I believe this to be true, and consequently I would refer to Sequoia couttsiæ the numerous specimens figured by Lesquereux as Glyptostrobus ungeri,[2] from the Green River group of Florissant, Colorado. I am also of the opinion that the specimens from the Fort Union group, at the mouth of the Yellowstone, described by Newberry[3] under the name of Glyptostrobus europæus Brongt., should be placed under Sequoia couttsiæ. I have never seen any of these specimens, however, and base this conclusion on the figures. I have seen a number of specimens from near the same place, collected in later years, and they seem to bear out this conclusion. Some of the material from the so-called Laramie of Canada also appears to be properly referable to this species. The whole subject needs thorough revision, with specimens at hand from all localities, and until this can be had no determinations can be regarded as final.

Habitat: Northeast side of Crescent Hill, opposite small pond, Yellowstone National Park; collected by F. H. Knowlton, August 2, 1888.

Monog. Brit. Eoc. Fl., Vol. II, Pt. I, Gymnospermæ, p. 29.
Cret. and Tert. Fl., p. 139, Pl. XXII, figs. 1–6a.
See Cret. and Tert., Pl. XI, figs. 6–8.

SEQUOIA LANGSDORFII (Brongt.) Heer.

PL. LXXXII, fig. 2.

Sequoia langsdorfii (Brongt.) Heer.: Fl. Tert. Helv., Vol. I, p. 54, Pl. XX, fig. 2; Pl. XXI, fig. 4.

This is by far the most abundant and widely distributed conifer found in the Yellowstone National Park, with the possible exception of *Sequoia notarisiá*, known only from the internal structure. It occurs in many places and in a variety of forms—that is to say, the branchlets and leaves are of various sizes, showing that they have come from many individuals and from different parts of the tree. They are not of the same size and character as specimens from the Fort Union group near the mouth of the Yellowstone,[1] being rather smaller and not so spreading, but they are very much like the typical leaves figured by Heer[2] from Greenland.

In all cases, however, the attachment of the leaves appears to be characteristic of this species. In one exceptional case the cellular structure of the leaf could be made out. This agreed perfectly with one given by Heer (loc. cit., fig. 21).

In one or two cases male aments were observed which much resemble those figured by Heer (loc. cit., fig. 19).

Habitat: Fossil Forest Ridge, beds Nos. 4, 6, and 7: collected by Lester F. Ward and F. H. Knowlton, August 16–20, 1887. Cliff west of Fossil Forest Ridge, altitude about 7,900 feet: collected by Lester F. Ward and F. H. Knowlton, August 15, 1887. East bank of Lamar River, between Cache and Calfee creeks: collected by F. H. Knowlton, August 21, 1888. Southeast side of hill above Lost Creek, bed No. 4: collected by F. H. Knowlton, August 9, 1888. Yancey fossil trees: collected by F. H. Knowlton, August, 1888. South end of Crescent Hill, about 300 feet above main wagon road, bed 6 feet below "Platanus bed:" collected by F. H. Knowlton, August 9, 1888. Northeast side of Crescent Hill, opposite pond: collected by F. H. Knowlton, August 2, 1888. Yellowstone below mouth of Elk Creek, bottom of bluff: collected by F. H. Knowlton, August 29, 1888. Also obtained by Mr. Arnold Hague (September 4, 1897) from

[1] Cf. Newberry: Il. Cret. and Tert. Fl., Pl. XI, fig. 4.
[2] Fl. Foss. Arct., Vol. I, Pl. II, figs. 2–22.

the south side of Stinkingwater Valley, at a high bluff east of the mouth of Crag Creek, Wyoming.

SEQUOIA, cones of.

Pl. LXXXI. fig. 8; Pl. LXXXII. figs. 6, 7.

The specimens figured are fairly representative of these organisms. They are quite fragmentary, yet appear to be cones. They are found in the same beds with *Sequoia langsdorfii*, but not in actual connection with that species, and I have preferred to keep them distinct, at least for the present.

Habitat: Fossil Forest, beds Nos. 5 and 6; collected by Lester F. Ward and F. H. Knowlton, August, 1887.

TYPHACEÆ.

PHRAGMITES ? LATISSIMA n. sp.

Pl. LXXXIII. fig. 5.

Leaf very broad; striæ fine, close together.

The fragment figured represents all that has been collected of this form. It is, of course, quite insufficient for proper diagnosis, yet it seems to be different from anything hitherto described from that region. It is certainly quite unlike anything found in the Yellowstone National Park.

It must have been a very large leaf, for the fragment is over 3 cm. broad, and it was probably a thick leaf. The striæ are very fine, straight, and close together. It differs in size and fineness of striæ from *P. alaskana*, to which it seems to be most nearly related.

I have given it a new name with great reluctance, for it is too fragmentary to found a new species on, but for the present it may remain as above.

Habitat: Northeast side of Crescent Hill, Yellowstone National Park; collected by F. H. Knowlton and G. E. Culver, August, 1888.

SPARGANIACEÆ.

SPARGANIUM STYGIUM Heer.

Sparganium stygium Heer. Cf. Ward: Types of the Laramie Fl., p. 18, Pl. III. figs. 6, 7.

These specimens do not agree in all particulars either with those figured by Professor Ward or with the types as shown by Heer. They are quite obscure, but in all probability they are identical with Heer's form.

Habitat: Yellowstone River, one-half mile below the mouth of Elk Creek, Yellowstone National Park; collected by F. H. Knowlton, August, 1888.

CYPERACEÆ.

CYPERACITES ANGUSTIOR (Al. Br.) Schimper.

Cyperacites angustior (Al. Br.) Schimp.: Pal. vég. Vol. II, p. 414, 1870.

Cyperites angustior Al. Br. Lesquereux: Ann. Rept. U. S. Geol. and Geog. Surv. Terr., 1872 (1873), p. 405.

This species was identified by Lesquereux, but the material can not now be found.

Habitat: "Elk Creek, near Yellowstone River; collected by A. C. Peale, Joseph Savage, and O. C. Sloane."

CYPERACITES GIGANTEUS n. sp.

Pl. LXXXII, fig. 10.

Leaves large, thick, with strong midvein and numerous close nerves.

This species, although fragmentary, is represented by several leaves and stems. The largest is 18 cm. in length and about 7 mm. in width. It has a well-defined midnerve or vein and numerous close veins. It was evidently of very firm texture.

Habitat: Yellowstone River, one-half mile below mouth of Elk Creek, at base of bluff; collected by F. H. Knowlton, August, 1888.

CYPERACITES? sp.

Pl. LXXXIII, fig. 4.

This fragment is all of this species, whatever it may be, that has thus far been found. It was at least 2.5 cm broad and had a well-marked keel. The veins are strong, about 1 mm. apart, with a fine intermediate vein. There is altogether too little of it to venture a specific description or determination.

Habitat: Cliff west of Fossil Forest Ridge; collected by Ward and Knowlton, August 15, 1887.

CYPERACITES? sp.

PL. LXXXIII, fig. 6.

The fragment figured is the only specimen of this kind in the collections, and it is too imperfect to be of much value. It is about 3 mm. broad, and appears to be two-ribbed—that is, two of the ribs are slightly stronger than the others. The fossil is rather faint, but there seems to be only one slender rib between the strong ones.

With so small a fragment it is impossible to be certain even of the generic reference, but in this respect it seems to agree with forms that have been referred to Cyperacites. It is left for future material to determine its status.

Habitat: Northeast side of Crescent Hill, Yellowstone National Park; collected by F. H. Knowlton and G. E. Culver, August, 1888.

SMILACEÆ.

SMILAX LAMARENSIS n. sp.

PL. CXXI, figs. 3, 4.

Leaves large, of firm texture, ovate or ovate-oblong, rounded truncate at base, abruptly and obtusely acuminate at apex, margin entire; 5-nerved or obscurely 5-nerved; petiole splitting into 5 equal branches, the middle of which passes straight to the apex, the others bending out to half the distance between midrib and margin, then curving toward and entering the tip; the lateral strong nerves branch on the outside, beginning first at the base of the blade, this branch sometimes reaching up to or near the apex or becoming lost near the middle; sometimes there are lateral strong nerves with 5 or 6 branches on the outside, all of which unite in broad loops to produce a false fifth nerve.

This species is represented by several large, finely preserved examples. The largest, represented in fig. 4, was 9 cm. broad and at least 14 cm. in length, and the smallest (see fig. 3) about 4 cm. broad and probably about 8 cm. in length.

In outline the leaves are broadly ovate or ovate-oblong, with a rounded truncate or cordate base. The apex in some cases appears to be long and rather obtusely acuminate, in other cases quite abruptly and obtusely acuminate.

The nervation is strongly marked. The petiole splits at the very base of the leaf, almost outside of the blade, into three equal divisions, one of which, the middle one, answering to the midrib, is straight and ends in the apex. The others arch out regularly and pass around and enter the apex also. Each lateral strong nerve or rib branches on the outside, the lowest branch, which originates just inside the lower margin of the blade, sometimes passing up and entering the apex with the other, at other times being lost below the middle of the blade. In still other instances there are 5 or 6 branches in the outside of the lateral ribs which join by a broad loop, forming a false nerve. In all cases this outside interrupted nerve is much lighter than the others.

Smilax appears to be rarely found fossil in North America, as only 5 species have been detected. Of these, 3 are from the Dakota group, and 1 each from the Laramie and Miocene. None of them is at all closely related to the one under discussion.

Smilax lamarensis seems to be closely related to a number of living forms. Thus the smaller rounded leaves are quite like *S. rotundifolia* L., both in shape and nervation, while the larger forms are hardly to be separated from *S. pseudochina* L. It is certainly closer to living American forms than any heretofore described from this country.

Habitat: East bank of Lamar River, between Cache and Calfee creeks; collected by F. H. Knowlton and G. E. Culver, August, 1888. Fossil Forest Ridge, bed No. 6, "Platanus bed," 1 large specimen; collected by Lester F. Ward and F. H. Knowlton, August, 1887.

MUSACEÆ.

MUSOPHYLLUM COMPLICATUM LX.

Pl. LXXXIII, fig. 1.

Musophyllum complicatum Lx.: Tert. Fl., p. 96, Pl. XV, figs. 4–6.

This species was described by Lesquereux from a "shale over a thin bed of coal, 8 miles southeast of Green River Station, Wyoming," in what he at first regarded as the Washaki group, but which he later[1] decided was

---

the true Green River group. This locality has not since been visited, and in fact can not now be satisfactorily located. It is more than probable, however, that the former determination of horizon is correct.

So far as I now know, this is the second time this species has ever been found. It is represented by 5 or 6 fairly well preserved specimens, which agree perfectly with Lesquereux's specimens and figures.

On one of the specimens there are a number of thick stems or stipes. They are longitudinally striate as described by Lesquereux, and bear only fragments of the leaves preserved. In the specimen figured we have a narrow leaf preserved almost entirely. It is about 5 cm. broad and 7 cm. long, as preserved, with perfectly entire margins. In still another specimen the stipe, with portions of the lamina attached, is fully 20 cm. long. There is no evidence from these specimens of the leaves having been as broad as described in some of the original specimens, but Lesquereux also speaks of narrow-leaved forms.

Habitat: Northeast side of Crescent Hill, opposite small pond, Yellowstone National Park; collected by F. H. Knowlton, August 2, 1888.

JUGLANDACEÆ.

JUGLANS CALIFORNICA LX.

Juglans californica Lx.: Foss. Fl. Aurif. Gray. Deposits. Mem. Mus. Comp. Zoöl., Vol. VI, No. 2, 1878, p. 35, Pl. IX, fig. 14; Pl. X, figs. 2, 3.

There are 2 specimens which I refer to this species. They are of the same size and nervation as the leaf shown in fig. 2, Pl. X, of Lesquereux's paper.

Habitat: Fossil Forest Ridge, bed No. 6, "Platanus bed;" collected by Ward and Knowlton, August 19, 1887. East bank of Lamar River, between Cache and Calfee creeks; collected by F. H. Knowlton and G. E. Culver, August 24, 1888.

JUGLANS RUGOSA LX.

Juglans rugosa Lx.: Cf. Tert. Fl., p. 286, Pl. LV, figs. 1-9.

The collections contain a number of very perfectly preserved specimens that undoubtedly belong to this species. Besides these there are a number of fragments that probably belong to it.

Habitat: Fossil Forest Ridge, beds Nos. 3, 5, 6, 7; The Thunderer;

Hill above Lost Creek, bed No. 4; Crescent Hill, 6 feet below Platanus bed; Yellowstone River nearly opposite Hellroaring Creek; collected by Ward and Knowlton, 1887, 1888.

JUGLANS SCHIMPERI LX.

Juglans schimperi Lx.; Tert. Fl., p. 287. Pl. LVI. figs. 5-10.

The collection contains a number of well-preserved examples that seem without doubt to belong to this species. The type is described by Lesquereux as having the margins slightly undulate. The Park specimens are slightly undulate, and also very slightly toothed. The nervation is absolutely the same in both.

Habitat: Andesitic breccia near gulch northwest of peak west of Dunraven; collected by Joseph P. Iddings, September 12, 1883.

JUGLANS LAURIFOLIA n. sp.

Pl. LXXXIII. figs. 2, 3.

Leaves large, membranaceous, lanceolate or ovate-lanceolate, slightly unequal-sided, narrowed to a wedge-shaped base and rounded to an acuminate apex, margins remotely and slightly denticulate; midrib thick below, becoming much thinner in the upper part; secondaries about 10 pairs, thin, alternate, emerging at an angle of 45° or 50°, camptodrome, much curving, ascending near the margin for some distance, where they arch by numerous loops; intermediate secondaries occasional; nervilles very irregular, much branched; finer nervation forming irregularly quadrangular areolae.

This species is represented by a number of well-preserved leaves. They are from 9 to 15 cm. long and 3 to 5 cm. wide, and are somewhat unequal-sided, the difference in the width of the sides being about 4 mm. This species is marked by its large size, remotely denticulate margin, and strong nervation.

Among fossil species, *Juglans laurifolia* is somewhat related to *J. egregia* Lx.,[1] from the Auriferous gravels of California. The leaves of the latter species differ, however, in being obovate-lanceolate, with the broadest part above the middle, and in having quite numerous, fine, sharp teeth.

[1] Mem. Mus. Comp. Zool., Vol. VI, p. 36, Pl. IX, fig. 12.

The nervation is somewhat similar, except that the secondaries are more numerous.

The relation between *J. laurelia* and *J. denticulata* Heer, from the Green River[1] group, Carbon, Wyoming, etc., while apparent, is much more remote than in the case of the California species. This species has very unequal-sided leaves, with large and more numerous teeth and more numerous arched secondaries.

Habitat: Hill above Yanceys and near the upright fossil trees; collected by F. H. Knowlton, August 28, 1888. Also found on southern spur of Chaos Mountain, altitude 10,100 feet; collected by F. P. King for Arnold Hague, August 11, 1897.

JUGLANS CRESCENTIA n. sp.

Pl. LXXXIV, fig. 8.

Leaflets large, of firm texture, lanceolate, narrowing to a long acuminate apex, truncate and slightly unequal-sided at base; margin perfectly entire, slightly undulate; midrib thin, straight, secondaries 15 or 18 pairs, alternate, at an angle of 35° to 45°, camptodrome, forking below the margin and joined to the one next above, forming a series of strong bows; intermediate secondaries frequent, about midway between two secondaries, thin and soon vanishing or rarely passing to the loop made by the secondaries; a series of small loops are produced by outside branches from the large bows which nearly or quite reach the margin; finer nervation dividing the space between the secondaries and intermediate secondaries by large quadrangular areolation.

This fine species is represented by a number of beautifully preserved specimens. The best is the one figured, which is 20 cm. in length and nearly 4.5 cm. in width. As stated in the diagnosis, it is truncate and slightly unequal-sided at the base and long acuminate at the apex. The margin is slightly undulate, but not otherwise cut or serrated. The secondaries are numerous, about 16 or 17 pairs, and strictly alternate. They fork at half or two-thirds of their length from the midrib and, uniting with the one next above, form a series of broad, strong loops well inside the margin. The finer nervation is very perfectly preserved, forming large but quite regularly quadrangular areolae.

Tert. Fl., p. 289, Pl. LVIII, fig. 4.

MON XXXII, PT II——44

Among described forms this has some resemblance in shape to *J. schimperi* Lx.,[1] from the Green River group of Colorado, but differs markedly in the forking of the secondaries. It is undoubtedly very close to *J. acuminata* Heer,[2] which in turn is hardly to be distinguished from the *J. rugosa* of Lesquereux. Additional material of all these will be necessary to settle the status of each.

Habitat: Northeast side of Crescent Hill; collected by F. H. Knowlton and G. E. Culver, August 2, 1888. Fossil Forest, bed No. 6; collected by Ward and Knowlton August, 1887.

HICORIA ANTIQUORUM (Newby.) Kn.

Hicoria antiquorum Newby., Kn.: Bull. U. S. Geol. Surv. No. 152, p. 117, 1898.
Carya antiquorum Newby.: Later Extinct Floras, p. 72; Ill. Cret. and Tert. Plants,
    Pl. XXIII, figs. 1–4. Lesquereux: Tert. Fl., p. 289, Pl. LVII, figs. 1–5.

The collection contains a number of somewhat fragmentary specimens, but the characteristic teeth and nervation suffice to enable their certain reference to this species.

Habitat: Fossil Forest Ridge, bed No. 5, "Platanus bed," collected by Ward and Knowlton, August, 1887.

HICORIA CRESCENTIA n. sp.

Pl. LXXXIV, fig. 5.

Leaflet thick and firm, elliptical-lanceolate, inequilateral; rather long wedge-shaped at base and apparently narrowed above to an acuminate apex; margin serrate, teeth small, sharp; midrib rather thick, straight; secondaries about 15 pairs, alternate, irregular, at obtuse angles, arching upward, rarely forked, craspedodrome, or suberaspedodrome, either arching near the margin and sending branches to the teeth, or dividing and sending weaker terminations into the teeth; intermediate secondaries occasional, short, and soon disappearing; nervilles numerous, mainly percurrent, approximately at right angles to the secondaries; finer nervation forming rather large quadrangular areolæ.

The specimen figured is the only one referred to this species, and unfortunately it lacks both base and apex. It is now about 7 cm. long and

2.8 cm broad. When entire, it must have been nearly or quite 10 cm in length.

This species is evidently related to *Carya? Heerii antiquorum* Newby., and it may possibly be an anomalous form of that species. It appears to differ essentially, however, in being very much smaller, less unequal-sided, and in having larger and less numerous teeth. The secondaries are much the same in both, except that they are more decidedly camptodrome in *C. antiquorum.*

It is also suggestive of *Juglans acuta* Ung., as identified by Ward[1] from the lower Yellowstone River.

Habitat: Northeast side of Crescent Hill, opposite small pond, at an altitude of about 7,500 feet; collected by F. H. Knowlton and G. E. Culver, July 27, 1888.

Hicoria culveri n. sp.

PL. LXXXIII, fig. 7.

Leaflets thin, slightly inequilateral, rather long obovate, narrowed from above the middle to a long wedge-shaped base, and upward to an acuminate apex; margin toothed from above the lower third, teeth flat, obtuse; midrib slender; secondaries about 10 pairs, alternate, irregular, camptodrome, arching upward and joining by a broad curve, with branches outside entering the teeth; intermediate secondaries occasional, joining the secondary below; nervilles very irregular, broken; finer nervation forming irregular meshes.

The fine specimen figured appears to be a terminal leaflet, as it is only slightly inequilateral. It is perfect, except at the apex. It is preserved for 8 cm, and was probably 9.5 or 10 cm long. It is 2.7 cm broad at the widest point, which is high above the middle. From this point it tapers regularly to the base and appears to pass rather abruptly to an acuminate apex.

This species has the same shape and arrangement as *Rhus bendirei* Lx.[2] from John Day Valley, Oregon, but differs in the serration of the margins and in the finer nervation. It seems likely that Lesquereux's species is a Hicoria rather than a Rhus.

[1] Types of the Laramie Fl., p. 33, Pl. XV, fig. 3.
[2] Proc. U. S. Nat. Mus., Vol. XI, 1888, p. 15, Pl. IX, fig. 2.

Among living species *H. culveri* appears closer to *H. ovata* (old *Carya alba* Nutt.), which has, however, sharper teeth and more regular nervilles. The secondaries and their branches entering the teeth are much the same in both.

Habitat: Yellowstone River, one-half mile below mouth of Elk Creek; collected by F. H. Knowlton and G. E. Culver, August, 1888.

MYRICACE.E.

MYRICA SCOTTII LX.

PL. LXXXIV, fig. 6.

Myrica scottii Lx.: Cret. and Tert. Fl., p. 147, Pl. XXXII, figs. 17, 18.

A single fairly well preserved leaf, that seems without question to belong to this species. It has been found before only at Florissant, Colorado.

Habitat: Yellowstone River, one-half mile below Elk Creek, at base of bluff; collected by F. H. Knowlton, August, 1888

MYRICA WARDII n. sp.

PL. LXXXIV, fig. 4.

Leaf of firm texture, lanceolate, long wedge-shaped at base, obtusely denticulate from some distance above the base; midrib thick; secondaries thin, rather scattered, alternate or subopposite, emerging at an angle of about 50°, arching evenly around and joining the one next above at a little distance from the margin, their union forming a continuous marginal line, from the outside of which small veins enter the obtuse teeth; nervilles thin, percurrent.

Unfortunately the fragment figured is the only specimen of this form, collected. This is 5.5 cm. long and about 12 mm. wide. When entire, it was probably 10 cm. or 12 cm. in length.

This species resembles more or less closely a number of described species, yet undoubtedly differs from all. Thus, in the matter of the intramarginal nerve it resembles *M. torreyi* Lx.[1] from the Montana and Laramie, but is much larger and has a greater number of parallel secondaries. In size and shape *M. wardii* is much closer to *M. polymorpha* Schimp.,[2] from

Tert. Fl., p. 129, Pl. XVI, figs. 3-10.
Cret. and Tert. Fl., p. 116, Pl. XXV, figs. 4, 5.

Florissant, Colorado. It differs essentially in having rather numerous secondaries, which parallel enter the somewhat sharper teeth. *M. tallis* Lx.,[1] also from Florissant, is evidently related, but has much larger, sharp teeth, and nervation as in *M. polymorpha* Lx.

Habitat: Fossil Forest Ridge, bed No. 5, "Salix bed," collected by Ward and Knowlton, August 19, 1887.

Myrica lamarensis n. sp.

Pl. LXXXIV, fig. 5.

Leaf firm in texture, narrowly lanceolate, narrowed below, apparently acuminate at apex; margin at some distance above the base, provided with numerous, small teeth; petiole short; midrib rather thick, straight; nervation pinnate, camptodrome, the secondaries joining and forming a thin line from inside the border; other nervation destroyed.

The little leaf has only the lower portion preserved. It is now 3.5 cm in length, and when entire was probably not less than 6 cm long. It is only 7 mm broad, and has a short petiole about 2 mm in length. The nervation is also destroyed, except the midrib and about a dozen secondaries, which are seen to be alternate. They arch and join, producing a marginal line just inside the margin.

This little leaf does not appear to have any very close relative in this country. Those approaching nearest, perhaps, are the narrow leaved species so common at Florissant, Colorado, such as *M. tallis* Lx.,[2] and *M. scottii* Lx.[3] They differ markedly, however, in the sharply serrate margin and craspedodrome nervation.

Myrica banksiaefolia Ung., as figured by Heer[4] from the Baltic Miocene, is perhaps nearer to our species. This has the narrowed base and similar teeth, but differs in the apex and in the nervation.

Habitat: East bank of Lamar River, between Cache and Calfee creeks; collected by F. H. Knowlton, August 21, 1888.

[1] Op. cit., p. 117, Pl. XXXII, figs. 41-46.
[2] Op. cit. and Text Fl. p. 117, Pl. XXXII, figs. 41-46.
[3] Loc. cit., p. 117, Pl. XXXII, figs. 47, 48.
[4] Misc. Balt. Fl., p. 67, Pl. XVIII, fig. 4.

SALICACEÆ.

POPULUS GLANDULIFERA Heer.

PL. LXXXIV, fig. 4.

Populus glandulifera Heer. Lesquereux; Cret. and Tert. Fl., p. 226, Pl. XLVII, figs. 3, 4. Ward. Types of the Laramie Fl., p. 19, Pl. IV, figs. 1-4.

The collection contains 3 specimens with very fine rounded or sharp teeth, that are referred without hesitation to this species, as figured by Lesquereux and Ward, from the typical Fort Union group of the lower Yellowstone. The best of these is figured.

Habitat: Yellowstone River, one-half mile below the mouth of Elk Creek; collected by F. H. Knowlton, August, 1888.

POPULUS SPECIOSA Ward.

PL. LXXXIV, fig. 3.

Populus speciosa Ward. Types of the Laramie Fl., p. 20, Pl. V, figs. 4-7.

The Park collection contains something over 50 specimens that are referred to this species. They are very perfectly preserved, and the petiole is in some cases 8 cm. in length. There can be no question as to their correct reference to the Fort Union species. I should also incline to refer to this species certain of the leaves described as *Populus subdiscolor* Ward and *P. acrobuncha* Ward, from the same beds. The very slight differences separating these forms seem hardly to be worthy of specific rank.

Habitat: Yellowstone River, one-half mile below mouth of Elk Creek; collected by F. H. Knowlton, August, 1888.

POPULUS ZADDACHI Heer.

Populus zaddachi Heer. Fl. Tert. Helv., Vol. III, p. 307. Lesquereux; Cret. and Tert. Fl., p. 158, Pl. XXXI, fig. 8.

A few specimens were obtained that must be referred to this species.

Habitat: Late acid breccia, Pyramid Peak; collected by Arnold Hague, 1897.

POPULUS XANTHOLITHENSIS n. sp.

Pl. LXXXV, figs. 1, 2.

Leaves large, nearly round in outline, but a little broader than long, truncate or slightly heart-shaped at base, rounded above; margin strongly toothed; teeth in lower portion simple and rounded, others, along the side of the blade, inclined to be double-toothed—that is, each large tooth has one or 2 smaller rounded projections; petiole long, slender; blade 7-nerved, central or midrib strong, straight, pair of ribs nearest the midrib originating at a little distance above the base of the blade, arching around and reaching the apex near together, other 2 pairs arising from the apex of the petiole and dividing equally the remaining portion of the blade; midrib with about 3 pairs of secondaries in the upper part, next pair of ribs with 4 or 5 branches on the outside, other ribs with numerous branches which anastomose, producing large irregular areas, with minute branches to the teeth; nervilles numerous, mainly broken, occasionally percurrent; finer nervation quadrangular.

The 2 beautiful specimens figured are the only ones referred to this species. Fig. 1, which lacks the upper portion, is 9.5 cm. broad and 7.5 cm. long. It has the petiole preserved for 5.5 cm. Fig. 2, which lacks a portion of one side and a fragment at the apex, is 8.5 cm. in length and 10 cm. in width. The petiole is not preserved.

The relation of this species is evidently with *Populus genatrix* Newby,[1] from the Lower Yellowstone. This is about the same size and much the same shape, except that it is more prolonged at the apex. The margin is described as having "rather small, appressed teeth," while those of *P. xantholithensis* are larger, sometimes double, and never appressed. The former species was described by Newberry as "5-nerved," although the figure shows it to be clearly 5-nerved. The present species is just as clearly 7-nerved. The midrib and its branches also differ markedly in the two forms.

This species has also some affinity with certain of the forms described by Professor Ward from the lower Yellowstone, but the relationship is not close.

[1] Later Ext. Fl., p.61; Ill. Cret. and Tert. Fl., Pl. XII, fig. 1.

Habitat: Yellowstone River, one-half mile below the mouth of Elk Creek, top of bluff; collected by F. H. Knowlton, August, 1888.

POPULUS DAPHNOGENOIDES Ward.

Pl. LXXXIV, fig. 2.

Populus daphnogenoides Ward: Types of the Laramie Fl., p. 20, Pl. VII, figs. 4–6.

The collection contains some 20 specimens that are referred to this species. They have the same general character, but are a little larger, with a more prolonged point and rather stronger nervation. There are no essential differences, however.

Habitat: Yellowstone River, one-half mile below the mouth of Elk Creek, top of bluff; collected by F. H. Knowlton, August, 1888.

POPULUS BALSAMOIDES Göpp.

Pl. LXXXVI, fig. 1.

Populus balsamoides Göpp, Lesquereux: Cret. and Tert. Fl., p. 218, Pl. LV, figs. 3–5.

The fragment figured is the only specimen of this species detected. It represents the basal portion of a medium-sized leaf, and agrees satisfactorily with the figures of this species as given by Lesquereux.

Habitat: Cliff west of Fossil Forest Ridge; collected by Ward and Knowlton, August 15, 1887.

POPULUS ? VIVARIA n. sp.

Pl. LXXXVI, fig. 2.

Leaf thick, roundish, or broadly elliptical in outline, toothed to near the base; teeth large, acute; nervation pinnate, camptodrome, or imperfectly craspedodrome; secondaries strong, opposite or subopposite, emerging at various angles, forking near the margin, the branches arching into bows and apparently sending branches from the outside to the teeth; nervilles obscure, but apparently percurrent.

This doubtful species rests on the single fragment figured. It was apparently about 10 cm long and 7 cm wide. It has some resemblance to certain species of Populus from the Fort Union group of the lower Yellowstone, as, for example, *P. genitrix* Ward,[1] but differs in the branching of

[1] Types of the Laramie Fl., p. 23, Pl. IX, fig. 1.

the lower large secondary, and somewhat in the teeth. The nerves are also craspedodrome in *Platanopsis*.

It is really too fragmentary and uncertain for identification, yet it differs from anything else found in the collection, and is simply named in a purely provisional manner, awaiting subsequent collections.

Habitat: West end of Fossil Forest Ridge; collected by Ward and Knowlton, August 15, 1887.

SALIX VARIANS Heer.

PL. LXXXV, fig. 5.

Salix varians Heer: Fl. Tert. Helv., Vol. II, Pl. LXV, figs. 1–5, 6–16.

The example figured certainly belongs to this species. It is the same shape, but a little larger, and has the same crose-dentate margin and the same midrib and general nervation.

Habitat: Lamar River, between Cache and Calfee creeks; collected by Knowlton and Culver, August 27, 1888.

SALIX ANGUSTA Al. Br.

Salix angusta Al. Br. Lesquereux: Tert. Fl., p. 168, Pl. XXII, figs. 4, 5; Cret. and Tert. Fl., pp. 157, 247, Pl. LV, fig. 6.

This species, originally described by Heer from the Swiss Tertiary, has been found by Lesquereux in the Green River group at Florissant, Colorado, and in the Miocene of Oregon. A number of doubtful fragments were reported from Spring Canyon in the Bozeman coal field, but they are too fragmentary to be of any value.

Habitat: Lamar River, between Cache and Calfee creeks; doubtful fragments; collected by F. H. Knowlton, August, 1888; also specimens No. 1967 of Hague's Park collection.

SALIX LAVATERI Heer.

Salix lavateri Heer: Fl. Tert. Helv., Vol. II, p. 28, Pl. LXVI, figs. 1–12; Fl. Foss. Alask., Pl. II, fig. 10. Lesquereux: Proc. U. S. Nat. Mus., Vol. XI, 1888, p. 35.

Habitat: South end Crescent Hill, bed 25 feet above "Platanus bed;" collected by F. H. Knowlton, August, 1888.

SALIX ELONGATA ? O. Web.

Salix elongata O. Web. Lesquereux: Tert. Fl., p. 169, Pl. XXII, figs. 6, 7.

A single quite well preserved specimen that seems to belong to this species. The nervation, however, is not well preserved, but as nearly as can be made out it may be referred to this form.

Habitat: Fossil Forest, lower stratum, No. 1221 of Hague's Yellowstone National Park collection; collected by Arnold Hague, September 24, 1884.

BETULACEÆ.

BETULA HOOINGSI n. sp.

Pl. LXXXVI, figs. 4, 5.

Leaves membranaceous, ovate, slightly unequal-sided, rather abruptly rounded to the base, more prolonged above; margin regularly toothed from near the base, teeth slightly unequal, a little hooked; nervation pinnate and craspedodrome; midrib well marked, straight; secondaries about 10 pairs, mainly alternate, occasionally opposite, arising at an angle of about 45°, straight or nearly so, terminating in the larger teeth, often with forks near the margin, all of which enter the other teeth; nervilles obscure, but apparently percurrent and at right angles to the secondaries; finer nervation not preserved.

This species is represented by 5 very perfect leaves, all of which are preserved on the same piece of matrix. The most perfect one figured is 8 cm. in length and 4.5 cm. wide, while the other is about 8 cm. long and less than 4 cm. wide. The petiole belonging to this specimen is 7 mm. in length.

This species somewhat resembles a number of described forms, as, for example, *Betula stevensoni* Lx.,[1] from Carbon, Wyoming, from which it differs somewhat in shape, number of pairs of secondaries, and in the more regularly serrate margin. *Betula elliptica* Sap. as identified by Lesquereux[2] from John Day Valley, Oregon, is perhaps closer, yet this differs in having only 6 or 7 pairs of secondaries and also in the teeth. *Betula prisco-dentata* Lx. from the same locality, has the same kind of teeth, but differs in size.

---

[1] Tert. Fl., p. 129, Pl. XVIII, figs. 4, 5.
[2] Cret. and Tert. Fl., p. 232, Pl. LI, fig. 6.

Besides these, belonging to the genus Betula, there are a number of others more or less resembling this leaf, e. g., *Alnus corpinoides* Lx.,[1] from Bridge Creek, Oregon, and *Celastrus ovatus* Ward, from the Fort Union group of Montana.

Among living species this appears to be closest to *Betula lutea* Michx., but even this is somewhat remote. In the future it may be thought best to place this fossil species under some other genus, for which, no doubt, reasons may be found, but for the present it seems best to place it in Betula.

Betula aquatilis Lx.[2] from the Auriferous gravels of California, is evidently closely related; but this differs in being much narrower, more wedge-shaped at base, and in having fewer and smaller teeth. *B. prisca* Ett., as figured by Ward[3] from the uppermost Fort Union, near the mouth of the Yellowstone River, is similar but much smaller.

This species is named in honor of Prof. J. P. Iddings, of the University of Chicago, who pointed out the locality at which it was first collected.

Habitat: Yellowstone River, one-half mile below the mouth of Elk Creek, top of bluff; collected by F. H. Knowlton, August, 1888. Hill above Lost Creek, bed No. 2; collected by F. H. Knowlton, August 5, 1888.

CORYLUS MACQUARRYI (Forbes) Heer.

Pl. LXXXVI, fig. 5.

Corylus macquarryi (Forbes) Heer; Urwelt d. Schweiz, p. 324, 1865.

The collection contains a single well-preserved specimen that is referred with some little hesitation to this well-known species. It has the proper shape, including the heart-shaped base and identical nervation, but differs slightly in the marginal dentation. In the typical form the margin has numerous rather small, sharp, upward-pointing teeth, while the one under consideration has fewer, rather blunt teeth. It may not be the same, but rather than make a new species I have referred it to *C. macquarryi*.

Habitat: Fossil Forest, middle stratum, Hague's Yellowstone National Park collection (No. 1220); collected by Arnold Hague, September 24, 1884.

Op. cit., p. 283, Pl. LI, fig. 5.
Mem. Mus. Comp. Zool., 1878, p. 2, Pl. I, figs. 2-4.
Types of the Laramie Fl., Pl. XIV, fig. 2.

FAGACEÆ.

FAGUS ANTIPOFII Abich.

Fagus antipofii Abich. Lesquereux: Ann. Rept. U. S. Geol. and Geog. Surv. Terr., 1872
    1873., p. 303.

Identified by Lesquereux, but not since observed.

Habitat: "Elk Creek, near Yellowstone River: A. C. Peale, Joseph
Savage, and O. C. Sloane."

FAGUS UNDULATA n. sp.

Pl. LXXXV, figs. 4, 5.

Leaves small, of very firm texture; elliptical with a broadly wedge-
shaped base and apparently obtuse apex; margin regularly undulate-
toothed, the teeth being regularly rounded and separated by similarly
rounded sinuses; midrib strong, straight; secondaries numerous, opposite,
parallel, unbranched, all entering the obtuse teeth; nervilles very numerous,
at right angles to the secondaries, usually broken and anastomosing,
although sometimes irregularly percurrent; finer nervation producing
small, irregularly quadrangular areolation.

This fine species is fortunately represented by several very perfectly
preserved examples, the two figured showing both the basal and apical portions They vary in length from 6 to 10 cm. and in width from 2.75 to 4
cm. The margins are very regularly undulate-toothed, the sinuses being
almost an exact reverse of the nearest teeth. The wedge-shaped base is,
however, without teeth for a short distance. The secondaries are at an
angle of about 45°. They are parallel and all enter the obtuse teeth. All
of the finer nervation is beautifully preserved and is seen to be irregularly
quadrangular.

This species does not approach closely to any living species known to
me. It is perhaps nearest to certain forms of the common American *F.
ferruginea* Ait., but the living form differs in being proportionately broader,
and when toothed has sharp teeth, quite like Castanea, and pointing upward.

Among the 80 or more fossil species that have been described from
various parts of the world, there are several that our species more or less
closely resembles. Of these, *F. dentata* Göpp., as identified by Heer[1] in the

Eocene beds at Atanekerdluk, Greenland, perhaps approaches most closely. They are, however, larger leaves, with coarser, more acute teeth and fewer strictly alternate secondaries. The leaves of *F. undulata* are quite unlike the type specimen of *F. dentata* as described by Göppert[?] from the Tertiary of Schossnitz, as indeed are the leaves doubtfully so identified by Heer.

The Yellowstone National Park leaves are also quite like leaves of *Fagus castaneæfolia* Ung., as figured by Heer[?] from the same beds. These were afterwards referred by Heer[?] to his *Castanea ungeri* on what seems to me to have been insufficient grounds. There is hardly any difference between these and leaves of *F. undulata*, except that the teeth are a little sharper. It is probable that they should be placed together, but as the status of Heer's plants is somewhat unsettled, I have preferred to keep them separated for the present.

Göppert has also described another species, *F. attenuata*, from Schossnitz, which is really quite close to *F. undulata*. It is about the same size and has rounded teeth, but there is uniformly a tooth between two of the teeth which are entered by the secondaries. In *F. undulata* every tooth is entered by a secondary.

Fagus antipofi Abich, as figured by Heer[?] from the so-called Miocene of Alaska, is not greatly unlike the species under consideration. It has the outline, parallel secondaries, and finer nervation, but not the same kind of teeth. There are a number of other species, as *F. atlantica* Ung., *F. feronæ* Ung., as figured by Lesquereux[?] from Elko, Nevada, etc., that resemble *F. undulata* in one or more particulars, but not by any means sufficiently for specific identity.

But among all fossil forms, two of the leaves described by Heer as *Castanea ungeri* from the supposed Miocene of Alaska, are undoubtedly nearest to the species under consideration. They are of about the same size and shape, but have teeth a little more acute. The secondaries are numerous, parallel, and enter the teeth as in *Fagus undulata*, but in origin

Tert. Fl. v. Schossnitz in Schlesien, Gædit., 1855, p. 18, Pl. V, fig. 11.
Fl. Foss. Arct., Vol. I, p. 106, Pl. X, fig. 7b; Pl. XLVI, figs. 1–5.
Loc. cit., Vol. II. Fl. Foss. Alask., p. 32.
Loc. cit., Pl. VII, fig. 4.
Cuba. Prot., Pl. XXVII, fig. 2.
Tert. Fl., p. 116, Pl. XIX, figs. 4–5.
Fl. Foss. Arct., Vol. II. Fl. Foss. Alask., Pl. V, figs. 1, 2.

they are uniformly alternate instead of opposite. The 2 leaves figured by Heer I regard as very doubtful. From their general facies they are much more likely to belong to Fagus than to Castanea. The other leaf figured with them[1] does not appear to be the same, and is probably a Castanea, although somewhat anomalous. They are found associated in the same beds with 2 species of Fagus, from which they are hardly to be separated.

Habitat: Bluff on Yellowstone River 1 mile below mouth of Elk Creek, and about same distance above mouth of Hellroaring Creek; collected by F. H. Knowlton, August 4, 1888.

CASTANEA PULCHELLA n. sp.

PL. LXXXVI, figs. 6-8; PL. LXXXVII, figs. 1-5.

Quercus drymeja Ung. Lesquereux; Cret. and Tert. Fl., p. 245, Pl. LIV, fig. 4.

Leaves of very thick, firm texture; long-lanceolate in outline, with wedge-shaped base and long, slender, acuminate apex; margin evenly and regularly toothed; teeth large and sharp, separated by prominent sinuses, or more obtuse with shallower sinuses; petiole long, slender; midrib strong, straight; secondaries very numerous, opposite or alternate, parallel, all, except two or three of the lowest, entering the teeth; nervilles well preserved, numerous, at right angles to the secondaries, mainly broken.

This fine species is represented by a very large series of specimens, nearly all in excellent state of preservation. They range in size from about 8 to 20 cm. in length and from 2 to 6 cm. in width, while the petiole in some cases is 3.5 cm. long and rather slender. They are lanceolate in outline, with a long wedge-shaped base, which is without teeth for some distance, and a very long slender apex provided with numerous strong teeth. The teeth of the margin are numerous and regular, in some cases, as in fig. 2 of PL. LXXXVII, being very large and sharp, while in others they are less prominent. They are, however, all sharp and upward pointing. The secondaries are numerous, parallel, and entering the teeth. The finer nervation is well preserved, the nervilles being numerous and mainly broken in crossing.

It is with some hesitation that these leaves are described as new to science. At first they were thought to be the same as the leaves from the Auriferous gravels of California referred by Lesquereux to *Castanea ungeri* Heer, but a careful study of the Yellowstone National Park material, comprising nearly 100 specimens, and of a fine collection from Independence Hill, Placer County, California, has convinced me that they are distinct, although closely related. The California species differ in having a shorter petiole, in the wedge-shaped base being destitute of teeth for a greater distance, in having serrate margins rather than Castanea-like teeth, and in having in general closer secondaries. This study has also brought out the fact that Lesquereux could hardly have been correct in identifying the California specimens with *Castanea ungeri* as figured by Heer from Alaska. As already stated under *Fagus undulata* (p. 700), it is more than probable that 2 of the leaves figured by Heer (loc. cit., Pl. VII, figs. 1, 2) should be restored to Fagus, and the other is certainly specifically distinct from the California leaves. The California specimens, as stated, differ also from the Yellowstone National Park species, and should probably be given a new name.

Lesquereux identified as *Quercus drymeja* Ung., a single leaf from Bridge Creek, Oregon, that must certainly be the same as *Castanea pulchella*. It is, for example, absolutely indistinguishable from fig. 7 of Pl. LXXXVI and fig. 2 of Pl. LXXXVII. A comparison of certain of the European figures of *Q. drymeja* makes it more than probable that it was not correctly identified among the Bridge Creek material. The leaf figured by Lesquereux is referred to *C. pulchella*, and *Q. drymeja* should be stricken from the west-coast flora, at least so far as it depends on this particular specimen.

It was at first thought best to separate the small leaves represented in fig. 7 of Pl. LXXXVI and figs. 2 and 3 of Pl. LXXXVII, as a distinct species, but the only difference is one of size, and in the large series at hand this breaks down. All gradations from the smallest to the largest may be found, which is quite in accord with the well-known differences in size of leaves to be found on living Castanea.

Habitat: Fossil Forest Ridge, Yellowstone National Park, bed No. 7, altitude about 7,250 feet; collected by Lester F. Ward and F. H. Knowlton, August 16–20, 1887. Yellowstone River, one-half mile below mouth of Elk

Cret. and Tert. Fl., p. 250, Pl. LII, figs. 1, 3-7.
U. Foss. Arct., Vol. II, Fl. Foss. Alask., p. 52, Pl. VII, figs. 4-5.
Cret. and Tert. Fl., p. 245, Pl. LIV, fig. 4.

Creek, bluff 300 feet above stream; collected by F. H. Knowlton, August 27, 1888. Junction Butte Fossil Forest, altitude about 7,450 feet; collected by Lester F. Ward and F. H. Knowlton, August 25, 1887.

QUERCUS GROSSIDENTATA n. sp.

Pl. LXXXVII, fig. 7.

Leaf large, coriaceous, broadly lanceolate (base destroyed), apex acuminate; margin strongly toothed, the teeth sharp, upward pointing; midrib perfectly straight; secondaries about 8 or 9 pairs, alternate, at an angle of 45°, craspedodrome, slightly arching upward, ending in the large teeth; nervilles strong, at right angles approximately to the midrib, mainly percurrent, but occasionally forked or broken, finer nervation not retained.

Unfortunately this fine species is represented by the single specimen figured, and this, it may be seen, lacks the basal portion. The part retained is 10 cm. long and 4.5 cm. wide. It was probably 14 or 15 cm. in length when perfect. It has the margin strongly toothed, the teeth with long, rounded or sharp points, each of which is entered by a secondary.

Habitat: Fossil Forest Ridge, bed No. 5; collected by Ward and Knowlton, August 19, 1887.

QUERCUS CONSIMILIS? Newby.

Pl. LXXXVII, fig. 6.

Quercus consimilis Newby.: Proc. U. S. Nat. Mus., Vol. V, p. 505, 1882 [1883].
Quercus breweri Lx.: Cret. and Tert. Fl., p. 246, Pl. LIV, figs. 5-8.

This is only a fragment of the base of a leaf. It does not agree absolutely with the figures of Lesquereux, but rather than make it a new species I have referred it provisionally as above.

Habitat: Yellowstone River, one-half mile below mouth of Elk Creek, at top of bluff; collected by F. H. Knowlton, August, 1888.

QUERCUS? MAGNIFOLIA n. sp.

Pl. LXXXVIII, fig. 1.

Leaf large, of firm texture, long, broadly obovate, narrowed to the base, rounded-obtuse at apex; margin at base entire, remainder of margin destroyed, but probably toothed or lobed; midrib thick, straight; secondaries about 18 pairs, alternate, at various angles, curving upward, apparently camptodrome; finer nervation entirely effaced.

The figured specimen is 19 cm. in length, and was probably at least 22 cm. in length when entire. It is about 7 cm. broad in the widest part, which is above the middle of the leaf. Unfortunately the margin, with the exception of a small portion near the base, is destroyed, and consequently it is impossible to properly characterize this leaf. There is, however, a little evidence to show that the margin was not entire for the whole distance, but this is too vague to be of much value.

I have referred this leaf provisionally to the genus Quercus, from its resemblance to certain living forms, but it will be necessary to see additional material before the correctness of this view can be tested.

Habitat: Yellowstone River, one-half mile below mouth of Elk Creek, at top of bluff; collected by F. H. Knowlton, August, 1888.

QUERCUS FURCINERVIS AMERICANA KN.

Pl. LXXXVIII, fig. 5.

Quercus furcinervis americana Kn.: Bull. U. S. Geol. Surv. No. 152, p. 192, 1898.
Quercus furcinervis Rossm.: Lesquereux: Cret. and Tert. Fl., p. 211, Pl. LIV, figs. 1, 2.

The specimen here figured is certainly the same as that figured by Lesquereux (loc. cit., Pl. LIV, fig. 1) for this species.

Habitat: Fossil Forest Ridge, bed No. 5; collected by Ward and Knowlton, August 19, 1887.

QUERCUS WEEDII n. sp.

Pl. LXXXVII, fig. 1.

Leaves membranaceous, ovate, rounded at base, acuminate at apex, margin strongly, irregularly toothed, teeth minutely spiny-pointed; nervation pinnate; midrib straight; secondaries about 8 pairs, alternate, at an angle of about 45°, flexuose, craspedodrome, entering the teeth or forking near the margin and the branches passing into the teeth, or with strong nervilles crossing between 2 secondaries and sending a branch to the intermediate teeth; nervilles numerous, strong, at various angles,

MON XXXII, PT II——45

percurrent or forked and broken; finer nervation beautifully preserved, forming quite regular, large areolæ.

This beautiful species is represented by a considerable number of more or less perfect leaves, the best of which is figured. This figured example is 11 cm. in length and nearly 6 cm. in width. Others are only 7 cm. long and 3.5 cm. wide. None larger than the one figured were obtained.

One of the marked features of this species is the number of teeth, there being quite regularly twice as many as the number of secondaries. These intermediate teeth are usually a little smaller than the others, and are supplied with a branch from the middle of a strong nerville, which crosses between 2 secondaries at some distance below the margin. This character is so constant and so peculiar that it may even be of generic value, but for the present the species may be retained in the genus Quercus.

This species has a more or less close resemblance to a number of described fossil forms. It is, for example, somewhat like *Quercus eburnisfolia* Lx.,[1] from Golden, Colorado, Black Buttes, Wyoming, etc. This is more wedge-shaped at base, has more irregular teeth, which are supplied by branches from the forking secondaries. *Quercus grœnlandica* Heer,[2] as figured from Spitzbergen also belongs to this group, but is a much larger leaf, with relatively smaller teeth and forked secondaries.

The very much larger leaf figured by Newberry as a young form or variety of *Platanus haydeni* Newby.,[3] and coming from the Fort Union beds at the mouth of the Yellowstone, also belongs near this group. It is impossible to see any generic, or even specific, differences between this figure of *P. haydeni* and Heer's figure above referred to of *Quercus grœnlandica*. They must certainly be the same.

All of the species mentioned seem to be very close to the one under consideration, but they differ constantly by the manner of the supply of nerves to the secondary teeth.

I have named this species in honor of Mr. Walter Harvey Weed, who collected the best specimen observed.

Habitat. Fossil Forest Ridge, middle stratum; collected by W. H. Weed. Bed No. 6, "Platanus bed;" collected by Ward and Knowlton, August 19, 1887.

[1] Tert. Fl., p. 159, Pl. XX, fig. 14.
[2] Fl. Foss. Arct., Vol. II, Misc. Fl. Spitzb., K. Vetensk. Akad. Handl., Vol. VIII, No. 7., Pl. XII, fig. 5.
[3] Illustrations Cret. and Tert. Fl., Pl. XXI.

QUERCUS sp.

Pl. LXXXIX, fig. 7.

This is a fragment of the base of what appears to have been a large, thick leaf. It has a thick midrib, and alternate, thin, parallel, straight secondaries, which arise at an angle of about 45°. None of the finer nervation is preserved.

This has some resemblance to the basal portion of what has been described as *Q. cabera* (p. 708), but it was several times larger than this and lacks the marginal toothing. They come from the same beds.

Habitat: Yellowstone River, one-half mile below mouth of Elk Creek, at base of bluff; collected by F. H. Knowlton, August, 1888.

QUERCUS OLAFSENI Heer.

Quercus olafseni Heer: Fl. Foss. Arct., Vol. I, p. 109, Pl. XLVI, fig. 10. Lesquereux, Cret. and Tert. Fl., p. 224, Pl. XLVIII, fig. 4; p. 295, Pl. LIV, fig. 3.

There are a number of examples of this species, some of which are very well preserved. They are, with the exception of some minor details, identical with the figures given by Heer and Lesquereux. Thus only occasionally are they doubly dentate, and the secondaries rarely branch. They are undoubtedly the same as the leaves figured by Lesquereux.

This species was reported by Lesquereux from the Bad Lands of Dakota (Fort Union group), and from Table Mountain, California (Miocene?).

Habitat: Yellowstone River, one-half mile below the mouth of Elk Creek, top of bluff; collected by F. H. Knowlton, August 27, 1888.

QUERCUS YANCEYI n. sp.

Pl. LXXXIX, fig. 2.

Leaf of firm texture, broadly lanceolate, somewhat wedge-shaped at base and acuminate at apex, with undulate toothed margin; midrib strong, straight; secondaries 9 or 10 pairs, alternate, remote, emerging at a low angle, curving upward, the lower ones arching along the border, upper ones entering the teeth or often arching along and joining the one next above; nervilles few, irregular, broken; finer nervation forming irregular areolæ.

The leaf figured is the only one of this species observed. It is 10 cm. long and 17 mm. wide in the broadest part, which is about the middle. The entire leaf, with the exception of a fragment at the base, is preserved. The margin is entire for the lower third, then it is undulate-toothed, the teeth being rounded or rarely with a minute sharp point. In the lower part the secondaries arch and join, while those above either enter the teeth or join by a long loop and send a branch from the outside to the teeth. The finer nervation is well preserved, forming irregular meshes.

This beautiful species does not seem to be very closely related to any fossil oak with which I am familiar. Perhaps its nearest relative is *Q. lauriifolia* Newby.,[1] from "burned shales, over lignite beds, Fort Berthold, Dakota," the age of which is not well indicated, but is certainly Tertiary. It has only very faintly undulate-toothed margins, and the secondaries are at a more acute angle than in the one under discussion.

I have named this species in honor of Mr. John Yancey, proprietor of the stage station, and the namer of the fossil forest near by.

Habitat: Yancey Fossil Forest; slope near the standing trunks; collected by Lester F. Ward and F. H. Knowlton, August 10, 1887.

QUERCUS CULVERI n. sp.

Pl. LXXXVII, fig. 5.

Leaf small, of thick, firm texture, approximately oblong in general outline, obscurely 5-lobed; margin strongly, irregularly toothed; teeth obtuse or rounded, pointing outward, separated by broad, shallow sinuses; petiole slender; midrib rather slender, nearly straight; secondaries 6 pairs, subopposite, emerging at an angle of about 45°, or the 2 lower pairs only 25°, all nearly straight and entering the teeth; third pair of secondaries longest, entering the lateral lobes, with branches on the lower side which pass to smaller teeth; nervilles strong, apparently broken; finer nervation not preserved.

This beautiful species is represented only by the specimen figured. It is a small leaf, being about 7.5 cm. long, including the petiole, which does not seem to be entirely preserved, and 5 cm. broad between the lateral

[1] Proc. U. S. Nat. Mus., 1882, p. 505; Plates issued, Pl. LIX, fig. 5, 5.

lobes. As already stated, the leaf is obscurely oblong in general outline, being slightly wedge-shaped at base and having the strongest teeth of lateral lobes at about two-thirds of the distance from the base. The apex is not preserved, but judging from the contour it must have been rather obtuse. The teeth of the margin are also rather obtuse. The nervation is strongly craspedodrome, the secondaries or branches all entering the teeth. Only a few nervilles are preserved and these appear broken. None of the ultimate nervation has been preserved.

This species is quite unlike any American fossil species with which I am familiar. Among living species it approaches quite closely to occasional leaves of *Q. prinoides* Willd. of the eastern United States. The living leaves incline to be more wedge-shaped at base and to have stronger teeth separated by deeper sinuses. It is hardly probable that the resemblance is close enough to warrant the assumption that *Q. prinoides* has actually descended from this fossil form.

I take pleasure in having named this fine species in honor of Prof. George E. Culver, some time professor of geology in the University of South Dakota, who assisted me in making the collection of plants in the Yellowstone National Park.

Habitat: Bank of Yellowstone River, one-half mile below the mouth of Elk Creek; top of bluff 300 feet above stream, in white, coarse-grained tuff; collected August 28, 1888, by F. H. Knowlton and G. E. Culver.

QUERCUS HESPERIA n. sp.

Leaf of firm texture, broadly lanceolate in outline, passing from about the middle down into a long wedge-shaped base, rather abruptly pointed at apex; margin with few (8 to 10) strong, sharp, upward-pointing teeth; midrib strong; secondaries 10 to 12 pairs, alternate, straight or slightly curving, ending directly in the teeth; intermediate secondaries frequent, about midway between the secondaries, disappearing about halfway between midrib and margin; nervilles irregular, producing large, coarse areolation; finer nervation similar.

The specimen upon which this species is founded is nearly perfect, lacking only the tip. It is 6 cm. long and a little more than 2 cm. wide. The lower half of the leaf is regularly wedge-shaped and the upper portion is apparently rather abruptly pointed.   The teeth are strong and sharp-pointed, with rounded sinuses.

This species seems to be allied to *Q. bowcainai* Lx.,[1] from California, but differs essentially in having much larger, sharper teeth and straight secondaries.   It is also allied to *Q. gowegi* which has undulate or slightly toothed margin and fewer, more curved secondaries.   It somewhat resembles a leaf that has been described as *Hicoria culteri*, which, however, differs in the teeth, and in having a camptodrome instead of a craspedodrome nervation.

Habitat. Yellowstone River, one-half mile below mouth of Elk Creek, at top of bluff; collected by F. H. Knowlton, August, 1888.

DRYOPYLLUM LONGIPETIOLATUM n. sp.

Pl. LXXXVIII, figs. 6, 7.

Leaves lanceolate, long wedge-shaped at base, long narrowly acuminate at apex, margin regularly undulate-toothed, the teeth sharp, upward pointing, separated by rather shallow sinuses; petiole very long, slender; midrib thick, straight; secondaries numerous, alternate, 12 pairs or more, at a low angle in the lower part, more acute above, slightly curving outward in passing to the margin, all ending in the teeth; nervilles at right angles to the secondaries, obscure but apparently mainly percurrent; finer nervation destroyed.

This species is represented by a number of specimens, none of which are complete in a single example, but by combining several a good idea of the species is given.   The length appears to have been about 20 cm. and the width in the middle 4 cm.   The petiole is long, being 2.5 cm., and possibly not all preserved.   In the larger leaves the secondaries are quite remote and distinctly alternate.   They arch slightly in passing to the teeth.

The leaves of this species were at first confounded with leaves of *Castanea pulchella*, with which they occur in the same beds, but they differ in the longer petiole, the smaller teeth, and in the irregular, arching secondaries, with an occasional intermediate secondary between   The teeth of the upper third of the leaf are also of a different character.

Among species of Dryophyllum this species has some resemblance to *D. aquamarum* Ward[1] from Black Buttes, Wyoming. The latter differs in being broadest below the middle, undulate or sinuate toothed, and in having more numerous, often camptodrome, secondaries. *D. subfalcatum* Lx.,[2] from Point of Rocks and Hodges Pass, Wyoming, also has some resemblance, but is much smaller, with more numerous close secondaries.

Habitat: Fossil Forest Ridge, Yellowstone National Park, bed No. 7, "Castanea bed;" collected by Lester F. Ward and F. H. Knowlton, August 16–20, 1887.

ULMACEÆ.

ULMUS PSEUDO-FULVA? LX.

PL. LXXXVIII, fig. 2.

Ulmus pseudo-fulva Lx.: Mem. Mus. Comp. Zoöl., Vol. VI, p. 16, Pl. IV, fig. 5, 1878.

The fragment figured is all that has been found of this form, and it is doubtfully referred to this species.

Habitat: Lamar River, between Cache and Calfee creeks, Yellowstone National Park; collected by F. H. Knowlton, August, 1888.

ULMUS MINIMA? WARD.

Ulmus minima Ward: Types of the Laramie Fl., p. 45, Pl. XXII, figs. 3, 4.

A single small broken specimen is referred doubtfully to this species. It is of about the same size, but has the secondaries at a little lower angle, and has the nervilles well preserved. They are strong and percurrent. The margin is toothed, but the teeth are not well preserved.

This leaf is found on the same piece of matrix with *Ficus tiliæfolia?* Al. Br.

Habitat: Mountain back of Yanceys, near the fossil trees; collected by F. H. Knowlton, August, 1888.

---

[1] Types of the Laramie Fl., p. 26, Pl. X, figs. 2–4.

[2] Cf. *D. bruneri* Ward, now referred to *D. subfalcatum*: Types of the Laramie Fl., p. 27, Pl. X, figs. 5–8.

ULMUS RHAMNIFOLIA? Ward.

Ulmus rhamnifolia Ward: Types of the Laramie Fl., p. 45, Pl. XXIII, fig. 5.

This species is also represented by a single, much broken leaf, with only a small portion of the margin preserved. It has the size and nervation of Professor Ward's species, and is with hardly any doubt the same.

Habitat: Yellowstone River, one-half mile below the mouth of Elk Creek, top of bluff; collected by F. H. Knowlton, August, 1888.

ULMUS, fruits of.

PL. LXXXVIII, figs. 3, 4.

As it is impossible to determine the species of Ulmus to which these fruits belong, or properly to characterize them, I have preferred to leave them unnamed specifically.

Habitat: Yellowstone River one-half mile below mouth of Elk Creek, top of bluff; collected by F. H. Knowlton, August, 1888.

PLANERA LONGIFOLIA Lx.

Planera longifolia Lx.: Tert. Fl., p. 189, Pl. XXVII, figs. 1-6; Cret. and Tert. Fl., p. 164, Pl. XXIX, figs. 1-13.

The collection contains some 40 more or less well-preserved examples of this species, which agree very well indeed with the various figures given by Lesquereux. A number are so well preserved that the finer nervation is retained. The nervilles are numerous, parallel, and mainly percurrent.

Habitat: Fossil Forest Ridge, bed No. 3 (30 specimens); bed No. 5 (10 specimens); collected by Ward and Knowlton, August, 1887.

URTICACEÆ.

FICUS DEFORMATA n. sp.

PL. XCI, fig. 2.

Leaf large, thick, long-obovate, slightly unequal-sided at base, abruptly rounded above to an obtuse apex and rather abruptly narrowed below; margin entire, conspicuously indented or deformed on one side, the margin of indentation rounded; midrib thick; secondaries thick, 10 or 12 pairs, alternate, emerging at an angle of 35° to 45°, curving upward, camptodrome; none of the finer nervation preserved.

This leaf is 15.5 cm. long and almost 8 cm. wide in the broadest part, which is high above the middle of the blade. It is long-obovate, obtuse at apex, and very broadly or obtusely wedge-shaped at base. The margin is entire except for a curious indentation on one side, which has probably resulted from an injury of some sort. This indentation passes nearly to the midrib and has rounded lobes. The secondaries adjacent to this are also distorted, being much curved. The finer nervation is not preserved.

It is possible that the fragment of the base of a leaf described and figured on Pl. LXXXIX, fig. 7, is the same as this species, but it is obviously impossible to be certain of this. It is also undoubtedly related to, and is possibly identical with, *F. asiminaefolia* Lx.[1] from Placer County, California. This has the same shape and nervation, but for obvious reasons it is best to keep them separate, at least until additional specimens can be obtained.

Habitat: Yellowstone River, one-half mile below mouth of Elk Creek, base of bluff; collected by F. H. Knowlton.

FICUS UNGERI LX.

PL. XCI, fig. 5.

Ficus ungeri Lx.: Tert. Fl., p. 195, Pl. XXX, fig. 5; Cret. and Tert. Fl., p. 163, Pl. XLIV, figs. 1-3.

Habitat: Yellowstone River, 1 mile below mouth of Elk Creek, west side, and about same distance above Hellroaring Creek; collected by F. H. Knowlton, August 4, 1888.

FICUS sp.

PL. LXXXIX, fig. 3.

This is too fragmentary to permit even the generic determination, but it seems to belong to Ficus. It consists of the base of a thick leaf having a thick midrib, with rather thin parallel secondaries and a short very thick petiole.

Habitat: Yellowstone River, one-half mile below mouth of Elk Creek, base of bluff; collected by F. H. Knowlton, August, 1888.

---

[1] Cret. and Tert. Fl., Pl. LVI, fig. 3.

FICUS SHASTENSIS? LX.

Ficus shastensis Lx.: Proc. U. S. Nat. Mus., Vol. XI, 1888, p. 28, Pl. XI, fig. 3.

This species was described by Lesquereux from Shasta County, California. It was said to be 6 cm. long and 3.5 cm. broad, and with a very thick petiole. The Park leaf is 8 cm. long and 4.5 cm. wide, and lacks the petiole. The nervation appears identical, but I have hesitated to make a positive identification on such scanty material.

Habitat: Lamar River, between Cache and Calfee creeks, on the same piece of matrix as *Salix angusta* and *Lapadium knowltusii*; collected by F. H. Knowlton, August 27, 1888.

FICUS SORDIDA LX.

Ficus sordida Lx.: Mem. Mus. Comp. Zoöl., Vol. VI, No. 2, 1878, p. 17, Pl. IV, figs. 6, 7.

A single fragment, representing the lower side of a leaf of about the same size and nervation as fig. 7 of Lesquereux's plate.

Habitat: Specimen Ridge, Fossil Forest, "Platanus bed;" collected by Ward and Alderson, August 25, 1887.

FICUS DENSIFOLIA n. sp.

Pl. LXXXIX, fig. 1; Pl. XC, figs. 1, 2; Pl. XCI, fig. 1.

Leaves large, very thick, unequal-sided, irregular long-obovate, broadest at or above the middle, obtuse above, narrowed below to a rounded truncate or slightly heart-shaped base; margin entire or very slightly undulate; petiole not preserved; midrib very thick, slightly flexuose; secondaries 8 or 9 pairs, lower opposite or subopposite, others alternate; lower secondaries thin, nearly at a right angle with midrib, others irregular, remote, at various angles, much arching upward, occasionally forked, all camptodrome, and joining by broad loops; middle secondaries sometimes branched on the outside, the branches joining by broad loops near the margin; nervilles strongly marked, mainly broken, producing by union large quadrangular areas; finer nervation producing irregular quadrangular areas.

This fine species differs markedly from all others obtained in the Yellowstone National Park, and is quite unlike any American form. The smaller leaves are 13 or 14 cm. long and 5 or 6 cm. broad, while the larger examples are fully 25 cm. long and nearly or quite 10 cm. broad. They are broadly obovate, being broadest usually above the middle. At base the leaves are narrowed into a small rounded, truncate, or even heart-shaped part. Above they appear rather abruptly narrowed into an obtuse apex.

The nervation is strongly marked. The midrib is very thick, as are most of the secondaries, especially in the middle, when they pass to the broad portion of the blade. They are then alternate, thick, and sometimes forked, and not rarely branched on the outside. The secondaries and their branches are arched and joined by broad bows.

Habitat: Southeast side of Crescent Hill, largest specimen (Pl. LXXXIX, fig. 1); Yellowstone River, one-half mile below mouth of Elk Creek, one peculiar, somewhat doubtful, specimen; also one from base of bluff (Pl. XC, fig. 1); hill above Lost Creek, typical specimen. All of the above collected by F. H. Knowlton, August, 1888. Specimen Ridge, Fossil Forest, opposite Slough Creek, "Platanus bed" and bed 100 feet above same; specimens numerous; collected by Lester F. Ward and E. C. Alderson, August 25, 1887. Fossil Forest Ridge, bed No. 5, "Salix bed," one broken specimen; collected by Ward and Knowlton, August 9, 1887.

FICUS HAGUEI n. sp.

Pl. XC, fig. 5.

Leaf thick, broad, rounded-ovate, apparently rounded and truncate at base and rather abruptly acuminate at apex; margin entire; midrib thick, perfectly straight; leaf palmately 5-ribbed from above the base, apparently a pair of thin secondaries originating from or near the base of the lamina, then a pair of very strong subalternate ribs or secondaries, at an angle of about 45°, which arch upward, above this pair, in the upper part of the blade, are 4 or 5 pairs of alternate thinner secondaries at a lower angle; all of the secondaries are joined some distance from the margin by a broad loop, with another series of smaller loops outside these, at least in the lower portion of the leaf; a number of irregular intermediate secondaries occur between the primary secondaries; nervilles thin, irregular.

The specimen figured is the only representative of this strongly characterized species. It unfortunately lacks both base and apex, but the portion preserved is 8 cm. long and nearly 8 cm. wide. There is, of course, no means of knowing the configuration of base and apex, but from all indications it is probable that the base was rounded-truncate and the apex abruptly acuminate. It is well characterized by secondaries, of which the lower prominent pair are strongest and arch up and join by a broad loop to the secondaries above, producing a palmately ribbed leaf.

I am uncertain as to the correctness of this generic reference, but it seems to approach closer to Ficus than any other. In any case, it is so well marked that it can be readily recognized. It does not appear closely related to any fossil species with which I am familiar, but among living species it has considerable resemblance to *F. nedosa* Tey. and Binn., and *F. prosa ra* R., both from British India.

The species is named in honor of the discoverer.

Habitat: Fossil Forest Ridge, middle stratum; collected by Arnold Hague, September 24, 1884.

FICUS TILIÆFOLIA ? Al. Br.

A fragment of the basal portion of a large leaf, apparently of this species. It is, for example, very much like the figure given by Lesquereux,[1] from the Auriferous gravels of California.

Habitat: Hill above Yanceys and near the fossil trees; collected by F. H. Knowlton, August, 1888.

FICUS ASIMINÆFOLIA La.

Ficus asiminæfolia Lx.; Cret. and Tert. Fl., p. 250, Pl. LVI, figs. 1-3.

A single deformed leaf, which agrees in nervation with this species and with the upper portion of another perfect leaf.

Habitat: Fossil Forest Ridge, bed No. 3, "Magnolia bed;" collected by Ward and Knowlton, August, 1887. Yellowstone River, one-half mile below mouth of Elk Creek, base of bluff; collected by F. H. Knowlton, August, 1888.

ARTOCARPUS ? QUERCOIDES n. sp.

Pl. XCII, fig. 4.

Leaf large, thick, 5 (7?)-lobed, lower lobes large, rounded; upper lateral lobes smaller, turning upward, of about the same size at apex as

Mem. Mus. Comp. Zool., Vol. VI, No. 2, p. 28, Pl. IV, fig. 8, 1878.

central or terminal lobe, all separated by broad, rounded sinuses; midrib very thick below and to the middle of the leaf, from which point it rapidly diminishes to the apex; secondaries numerous, alternate, at angle of 30 to 45°, about 4 in each lobe, except the small central lobe, the upper ones passing to the apex of the lobe, the other curving near the margin below it; short secondaries pass up to and arch along above the sinuses, occasionally in the upper part forking and passing on both sides; nervilles strong, percurrent, nearly at right angles to the secondaries; finer nervation not preserved.

The specimen figured is the only one obtained of this remarkable and highly characteristic leaf. It is not perfect, yet it appears to represent practically all of the leaf. The part preserved is 14 cm. long and 9.5 cm. broad between the upper lobes. It was probably at least 17 cm. in length, and if there were 7 lobes it was of course much larger. It was probably 12 to 14 cm. broad between the lower lobes. The width at the middle sinus is a little less than 5 cm. It is strongly 5-lobed, and, following the analogy of *Artocarpus lessigiana* (Lx.) Kn., may have been 7-lobed. There is, however, no evidence that it had more than 5 lobes. The lower lobe is 5.5 cm. wide at a distance of 1.5 cm. from the midrib, while the upper lateral lobe is fully 6 cm. wide at the same distance from the midrib. The extreme length of the upper lobe is less than 5 cm., the apex being curved around and up. The secondaries, as pointed out in the diagnosis, are about 4 in number in each lobe. They are about 1 cm. apart, the upper one only entering the apex of the lobe. The only trace of the finer nervation consists of a few strictly percurrent nervilles.

I am in doubt as to the proper generic reference of this leaf. When it was collected in the field, the conclusion was hastily formed that it was an oak, but the nervation is not at all that of this genus. It seems to have rather a moraceous character, but I have not been entirely successful in finding affinities. It has some resemblance to species of Ficus, but on the whole approaches closest to Artocarpus. Compared with living species it is of the *A. incisa* type, yet of course differs in marked peculiarities, having, for example, only five instead of many lobes. Among fossil species this undoubtedly approaches *A. lessigiana* (Lx.) Kn.[2] found in the Laramie and Denver formations of Colorado, Wyoming, etc. The Yellowstone leaf has much the shape and thick midrib of the other, but differs essentially in having 3 or 4 secondaries instead of 1 in each lobe. It is, however, a leaf sufficiently well characterized to permit it to be readily recognized, and if material is hereafter found that will throw additional light on its affinities, it can be easily transferred to its proper genus. For the present it may remain under Artocarpus.

Habitat: Yellowstone River, one-half mile below mouth of Elk Creek; collected by F. H. Knowlton, August, 1888.

MAGNOLIACEÆ.

Magnolia californica ? Lx.

Magnolia californica Lx.: Foss. Pl. Autif. Gravels, Mem. Mus. Comp. Zoöl., Vol. VI, No. 2, 1878, p. 25, Pl. VI, figs. 5–7.

A single specimen, of which only the upper part is preserved. It has, so far as can be made out, the shape and nervation of this species, but it is so much broken that its positive identification is not possible.

Habitat: Fossil Forest Ridge: Hague's No. 1960.

Magnolia spectabilis n. sp.

Pl. XCIII, figs. 1, 2.

Leaves very thick, coriaceous; broadly elliptical-lanceolate in outline, with regularly rounded base and rather abrupt obtusely acuminate apex; margin perfectly entire, not undulate; midrib thick, straight; secondaries about 18 or 20 pairs, alternate, regular and parallel or slightly irregular on emergence from the midrib, becoming parallel above; secondaries either forking near the margin or arching along and joining the one next above in a series of loops, with a series of smaller loops outside; intermediate secondaries usually numerous, sometimes passing nearly to the junction of the primary ones, or becoming lost at one-half or two-thirds of the distance from midrib to margin, irregular and not parallel to other secondaries; nervilles numerous, irregular, broken, approximately at right angles to the secondaries; finer nervation beautifully preserved, forming strongly marked quadrangular areolæ.

This fine species is represented by a large number of well-preserved specimens. The larger leaves are fully 20 cm. long and 7 or 8 cm. wide, and some of the smaller ones 12 or 15 cm. long and 4 to 6 cm. wide. The leaves are thick and leathery, and evidently belonged to an evergreen species.

It is altogether probable that the leaves obtained by Mr. W. H. Holmes in 1878 from Amethyst Mountain and identified by Lesquereux as *Magnolia lanceolata* Lx.,[1] really belong to this species. As nearly as can be made out from Holmes's description of the locality,[2] it is the same as that which afforded the specimens under discussion. But a careful comparison of these numerous leaves with the figures given by Lesquereux, as well as with specimens from the Auriferous gravels, makes it certain that they can not belong to *M. lanceolata*. *Magnolia spectabilis* differs in being broader, more rounded at base, with secondaries more curved and with numerous intermediate secondaries. A still greater point of difference is in the texture of the leaf. Of *M. lanceolata*, Lesquereux says: "This leaf is not coriaceous, rather of a thin substance," while *M. spectabilis* is thick and distinctly coriaceous or leathery. The finer nervation is not preserved in *M. lanceolata*, so it is not possible to compare that point.

From further evidence it appears that these identical specimens were again submitted to Lesquereux in 1887, and he then identified them with *M. inglefieldi* Heer,[3] a species that he has also reported from Lassen County, California, Green River group, etc. It is certainly much more closely related to this than to *M. lanceolata*, as may be seen from Heer's figures[4] and specimens identified with it from California. It is of the same shape and size as *M. spectabilis* and is described as being coriaceous, but it differs somewhat in having the secondaries more scattered, apex irregular, etc. The finer nervation also differs. They are undoubtedly close, but seem to be sufficiently distinct for specific separation.

Among living species the affinity of *M. spectabilis* is unquestionably with *M. grandiflora* L., or *M. fotida* Sargent, as it is now called. The size, outline, texture, and nervation are practically the same.

According to Sargent,[5] the direct ancestor of *Magnolia fotida* was

Mem. Mus. Comp. Zool., Vol. VI., p. 20, Pl. VI., fig. 1.
Twelfth Ann. Rept. U. S. Geol. and Geog. Surv. Terr., 1878-1883, Pt. II., p. 49.
Loc. cit., p. 20.
Cf. Proc. U. S. Nat. Mus., Vol. X, 1887, p. 46.
Fl. Foss. Arct., Vol. VII, 1883, p. 424, Pl. LXIX. fig. 1; Pl. LXXXV. fig. 5; Pl. LXXXVI. fig. 9.
Silva of North America, Vol. I, p. 3.

probably *M. inglefieldi* as exemplified from Greenland, the type locality. As already pointed out, this species is also the closest relative of *M. spectabilis*, which in turn is closely allied to *M. fatida*. It is possible that *M. inglefieldi*, the earliest arctic representative, was pushed down by the invading ice, occupying under a slight variation (*M. spectabilis*) the Yellowstone National Park, and surviving at the present day as *M. fatida*.

Habitat: Fossil Forest Ridge, bed No. 3, "Magnolia bed," collected by Ward and Knowlton, August, 1887.

MAGNOLIA MICROPHYLLA n. sp.

Leaf thick, elliptical-lanceolate in outline, with slightly undulate entire margin; midrib very thick, straight; secondaries 4 or 5 pairs, alternate, very irregular, at an angle of 30° to 45°, much curved upward, forking near the margin, the fork joined by the branch from the secondary next below; intermediate secondaries present, irregular; nervilles irregular; finer nervation obscure.

A single broken fragment is the only example of this species observed. The part preserved is 6 cm. long and 4 cm. wide.

This leaf was associated on the same piece of matrix with *M. spectabilis*, yet differs by the characters enumerated.

Habitat: Fossil Forest Ridge, bed No. 3, "Magnolia bed," collected by Ward and Knowlton, August, 1887.

MAGNOLIA CULVERI n. sp.

PL. NCH. fig. 5.

Leaf large, membranaceous, broadly ovate, truncate at base, obtusely pointed above; petiole short; midrib thin, straight; secondaries 6 or 7 pairs, alternate, at an angle of 40° or 45°, forking some distance below the margin, camptodrome by broad loops; intermediate secondaries occasional, soon lost in the middle area between the secondaries; nervilles numerous, irregular, thin, broken; finer nervation producing large irregularly quadrangular areas.

The specimen figured, which is that best preserved, is 14 cm. in length, including the petiole, but lacks the apex. It must have been some 15 cm. long when perfect. It is broadest just below the middle, where it is 8 cm. wide. The petiole is 1 cm. long and moderately thick.

This fine species belongs certainly to the genus *Magnolia*, as attested by the shape and the forking, camptodrome secondaries. It is, however, quite unlike any of the other species found in the Yellowstone National Park. Perhaps its closest relation is *M. californica* Ls.,[1] from the Chalk Bluffs of California. It is different in shape, being ovate instead of broadly oval, and has somewhat different secondaries. The large quadrangular finer nervation is similar in both.

It does not approach very closely to either of the living American species, being perhaps closest to *M. acuminata* L., the well-known cucumber tree.[2] The shape of the leaves is practically the same, but the nervation differs somewhat.

I have named this species in honor of Prof. George E. Culver, who assisted in making the collection from this place.

Habitat: East bank of Lamar River, between Cache and Calfee creeks; collected by F. H. Knowlton and George E. Culver, August, 1888.

MAGNOLIA ? POLLARDI n. sp.

Pl. LXXXI, figs. 9, 10.

Petals of firm texture, elliptical or elliptic-ovate in outline, narrowed below, rounded-obtuse above; nervation of numerous approximately parallel nerves about 2 mm. apart.

The best preserved of these 2 specimens (fig. 9) has 7.5 cm. in length preserved, and was probably fully 8.5 cm. in length when perfect. It is 3 cm. broad in the middle, and is narrowed at base to a point of attachment some 5 or 6 mm. broad. The upper point is unfortunately destroyed, but it seems probable, from the appearance of the margin and nerves, that it was obtuse. The nerves arise from the basal part and run approximately parallel, spreading slightly in the middle and converging toward the apex. In the middle these nerves are between 3 and 4 mm. apart, but in the apex they are separated by only about 2 mm. There is some evidence of intermediate nerves, or possibly cross veinlets, but these are so indistinct that a positive statement concerning them can not be made.

The other specimen (fig. 10) is a trifle over 5 cm. in length, but lacks both upper and lower parts. It was probably 6.5 or 7 cm. in length when

Mem. Mus. Comp. Zool., Vol. VI, No. 1, p. 25, Pl. VI, figs. 5, 7.
Cf. Sargent: Silva of N. A., Vol. I, Pls. IV, V.

perfect. It is exactly 3 cm. in width at the widest portion, which is a little above the middle. There is no indication of the form of the base, as it is destroyed. The apex was quite obviously obtuse. The nerves are less distinctly preserved than in the other specimen, but by careful search they can be made out as shown in the figure. Beyond these nothing can be made out.

It is with some hesitation that these specimens are described as petals of Magnolia. They were at first supposed to be spathe-like growths of some monocotyledonous plant, and their identification as Magnolia petals was first suggested by Mr. C. L. Pollard, of the United States National Museum, in whose honor I take pleasure in naming the species. The probability of their being petals of a large-flowered Magnolia is greatly strengthened by the fact that undoubted Magnolia leaves in abundance are found in the various beds of the Yellowstone National Park, whereas no monocotyledonous plant has been found to which these apparently could have belonged. There is a facies to the specimens that is difficult to describe and wholly impossible to show in a figure, which is very suggestive of Magnolia petals. The manner in which they curve and narrow on the rock, although this appearance may of course be only accidental, is very similar to the petals of certain large-flowered forms—such, for example, as *M. conspicua*. In any case they are distinctive forms that may be readily recognized, and, for the purposes of geologic correlation, are of undoubted value. Several botanists to whom the specimens have been submitted agree that their reference to Magnolia is fully warranted, and for the present at least they may be so considered.

Habitat. Yellowstone River, one-half mile below the mouth of Elk Creek (fig. 10); collected by F. H. Knowlton, August, 1888. Fossil Forest Ridge, opposite Slough Creek; collected by Lester F. Ward, August, 1887.

LAURACEÆ.

LAURUS PRIMIGENIA? Ung.

Pl. XCI, figs. 4, 5.

Laurus primigenia Ung. Cf. Ward: Types of the Laramie Fl., p. 47, Pl. XXIII, fig. 8.

The much broken specimens are the only ones of this species found. Their identification is open to doubt, yet they are obviously the same as the leaf figured by Professor Ward as this species from Carbon, Wyoming. More material will be necessary before its status can be fixed with certainty.

Habitat: Yellowstone River, half a mile below mouth of Elk Creek, foot of bluff; collected by F. H. Knowlton, August, 1888.

LAURUS PERDITA n. sp.

Pl. XCIV, figs. 4-5.

Leaves coriaceous, broadly lanceolate, wedge-shaped at base, obtusely acuminate at apex; margins entire, but very slightly undulate; petiole short, stout, midrib thick, straight; secondaries 7 or 8 pairs, alternate, camptodrome, arising at an angle of 40° or 45° and curving upward and arching along near the margin and forming numerous broad loops or bows; nervilles numerous, irregular, mainly forked, approximately at right angles to the midrib; finer nervation not preserved.

The collection contains a number of specimens of this species, none of them, however, quite perfect. They are about 15 cm. long and about 4.5 cm. broad. The 5 specimens figured show well the character of the species. They are broadly lanceolate, with a regularly narrowed base and apparently a rather obtuse apex. The secondaries are about 7 pairs, which arch much upward and along the borders. The nervilles are numerous, mainly at right angles to the midrib, and irregular and often broken.

This species has some resemblance to *Laurus grandis* Lx.[1] from the Auriferous gravels of California, differing in being smaller, narrower, and not so obtuse at apex. The resemblance is close enough, however, to make it reasonably certain that the 2 species are quite closely related.

Persea pseudocarolinensis Lx.[2] from Table Mountain, California, is somewhat similar, but differs in being broader, more obtuse, and in having finer nervation.

Habitat: Hill above Yanceys and near the standing fossil trees, collected by F. H. Knowlton, August 28, 1888. Near same locality; collected by George M. Wright, September 24, 1885.

[1] Cret. and Tert. Fl., p. 251, Pl. LVIII, figs. 1, 2.
[2] Mem. Mus. Comp. Zool., Vol. VI, No. 2, p. 19, Pl. VII, figs. 1, 2.

LAURUS MONTANA n. sp.

Pl. XCV, fig. 2.

Leaves large, evidently coriaceous, elliptical-lanceolate, narrowed gradually (?) to the petiole and (?) upward to an acuminate apex (?), slightly unequal-sided in the upper part; margin entire; midrib thin, straight; secondaries 5 or 6 pairs, alternate, the lower at a very acute angle, upper ones slightly less so, all, but especially the lower ones, with numerous branches on the outside, which join and form broad loops just inside the margin; nervilles strong, percurrent, approximately at right angles to the secondaries; ultimate nervation not preserved.

The leaf by which this fine species is represented unfortunately lacks both base and apex, but is otherwise well preserved. It is 10 cm. long as now preserved, and was, when entire, probably at least 14 cm. in length. The width is 5.3 cm. As stated, it is a little (3 mm.) wider on one side of the midrib than the other, making it slightly unequal-sided. The nervation is peculiar, consisting of about 5 pairs of secondaries, of which the lower, on the narrower side of the leaf, begins well toward the base and passes up to the middle of the blade, with numerous branches on the outside at right angles to the midrib. The lower secondary on the broad side of the leaf is very thin and short, and anastomoses with a branch from the lower portion of the second secondary. This latter is strong, and passes above the middle of the leaf, and has only 4 or 5 branches on the outside, all being at an acute angle with the midrib. The other secondaries have 4 or more branches on outside, and also a number of strong nervilles.

This species appears to be related to some of the forms figured by Lesquereux as *Laurus qualis*,[1] from California, and may possibly be an anomalous form of this species. It is larger, more rounded, slightly unequal-sided, and has quite different nervation. It also resembles *L. californica* Lx.,[2] from the same place.

Habitat: Yellowstone River, one-half mile below mouth of Elk Creek, base of bluff; collected by F. H. Knowlton, August, 1888.

Cret. and Tert. Fl., Pl. LVIII, fig. 3.
Op. cit., Pl. LVIII, fig. 8.

LAURUS PRINCEPS Heer.

Pl. XCV, fig. 5.

Laurus princeps Heer.: Lesquereux, Cret. and Tert. Fl., p. 250, Pl. LVIII, fig. 2.

The fine leaf shown in the plate is absolutely perfect. It has the same size, shape, and nervation as fig. 2 of Lesquereux's plate (loc. cit.).

Habitat: Yellowstone River, one-half mile below mouth of Elk Creek, base of bluff; collected by F. H. Knowlton, August, 1888.

LAURUS CALIFORNICA Lx.

Laurus californica Lx.: Cret. and Tert. Fl., p. 250, Pl. LVII, fig. 5, Pl. LVIII, figs. 6-8.

Habitat: Fossil Forest Ridge, beds Nos. 3, 5, and 6; Specimen Ridge, Fossil Forest, opposite Slough Creek; collected by Ward and Knowlton, August, 1887. Northeast side of Crescent Hill, opposite small pond, altitude 7,500 feet; collected August 2, 1888, by F. H. Knowlton and G. E. Culver.

LAURUS GRANDIS Lx.

Pl. XCIII, fig. 5; Pl. XCV, fig. 4.

Laurus grandis Lx.: Cret. and Tert. Fl., p. 251, Pl. LVIII, figs. 1-5.

Habitat: Fossil Forest Ridge, beds Nos. 3, 5, and 7; collected by Ward and Knowlton, August, 1887. Specimen Ridge, Fossil Forest, head of Crystal Creek; collected by Ward and Alderson, August 25, 1887. Hill above Lost Creek; collected by George M. Wright, September 24, 1885.

PERSEA PSEUDO-CAROLINENSIS Lx.

Pl. XCV, fig. 1.

Persea pseudo-carolinensis Lx.: Auriferous Gravels of California, Mem. Mus. Comp. Zoöl., Vol. VI, No. 1, p. 19, Pl. VII, figs. 1, 2.

The specimen figured, which is the best one found, agrees closely with the figure of this species given by Lesquereux (loc. cit., fig. 1).

Habitat: Specimen Ridge, Fossil Forest, head of Crystal Creek; collected by Ward and Knowlton, August 25, 1887. East bank of Lamar River, between Cache and Caliee creeks; collected by F. H. Knowlton, August 21, 1888.

MALAPOENNA LAMARENSIS n. sp.

PLATE III, figs. 4, 5; PL. XCVI, fig. 5.

Leaves thick, coriaceous, ovate-oblong, tapering downward to a long wedge-shaped base and upward to an acuminate or obtusely acuminate apex; margin entire; midrib thick, straight; nervation pinnate, consisting of 2 pairs of opposite thick ribs or secondaries, of which the lower pair arise near the base and pass up for nearly half the length of the blade, while the other arise some distance up and pass nearly or quite to the apex; several pairs of small secondaries arise from the midrib in the extreme upper part of the blade; ribs with occasional branches on the outside; nervilles apparently percurrent.

This species is represented by several specimens, 3 of the best of which are figured. Unfortunately none of the specimens are perfect. The larger and best-preserved specimen has 9 cm. retained, and must have been 11 or 12 cm. in length when complete. This specimen is 4 cm. wide. Another example has 7 cm. of the upper portion preserved and is about 4.5 cm. wide. The small one figured is not quite 4 cm. in length and about 4.5 cm. in width.

Among living species *M. lamarensis* very much resembles *Tetranthera* (*Litsea*) *dealbata* R. Br. from Australia, and also approaches *Litsea glauca* Seib., from Japan—that is, it approaches these living species closely enough to make it certain that the generic reference is correct. Among the fossil species, *Tetranthera praecursoria* Lx.,[1] from the Bad Lands of Dakota, is quite suggestive. This species is somewhat obovate instead of ovate-oblong, and has about 4 pairs of secondaries, which do not differ in size as they do in *M. lamarensis*.

Habitat: East bank of Lamar River, between Cache and Calfee creeks; collected by F. H. Knowlton, August 24, 1888. Yellowstone River, one-half mile below mouth of Elk Creek; collected by F. H. Knowlton, August 27, 1888.

LITSEA CUNEATA n. sp.

PL. XCII, figs. 2–4.

Leaf membranaceous, broadly lanceolate, wedge-shaped at base and narrowed in about the same manner at apex; midrib very thick, straight;

Cret. and Tert. Fl., p. 228, Pl. XLVIII, fig. 2.

secondaries at a very acute angle, craspedodrome, alternate, lower pair thinnest, those above much thicker, branching on the outside, branches at an acute angle, craspedodrome; intermediate secondaries several, generally lost in the space between the secondaries; nervilles strong, at various angles, mainly percurrent; finer nervation irregular.

No perfect example of this species has been found, the fragments figured being all that we have to represent it. The specimen showing the wedge-shaped base is only 5 cm. long, but was probably 10 or 12 cm. in length when perfect. It is 4 cm. wide. The larger of the others is the wedge-shaped apical portion, and is 6 cm. long, with the probability of its having been at least 12 cm. long. The small specimen was probably hardly more than 8 or 9 cm. in length when perfect. The upper portion appears to have more numerous secondaries than the lower part. They are also branched on the outside.

Habitat. Yellowstone River, 1 mile below the mouth of Elk Creek; collected by F. H. Knowlton, August, 1888.

CINNAMOMUM SPECTABILE Heer.

Pl. XCIV, fig. 6.

Cinnamomum spectabile Heer. Fl. Tert. Helv., Vol. II, p. 91, Pl. XCVI, figs. 4-8.

The leaf figured, which appears to be the only one obtained, differs slightly from the figures of the European form to which it is referred. The lower pair of secondaries, for example, are nearer the base of the leaf than in the figures given by Heer, but, granting the slight differences, I have hesitated to make it a new species.

Habitat: Tower Creek, Yellowstone National Park; collected by Arnold Hague (field No. 10566), August 16, 1883.

PLATANACEÆ.

PLATANUS GUILLELMÆ Göpp.

Pl. XCVI, fig. 4; Pl. XCVII, fig. 5.

This species is very abundant, being represented by over 125 more or less perfect specimens. Some of these—as, for example, the one figured— are particularly perfect. They differ somewhat in size, the average being about 7 or 8 cm. broad between the lobes and 8 or 9 cm. in length. An occasional one is 14 cm. broad and about the same length.

These leaves agree well with the usual description and figures of this species, especially as given by Lesquereux[1] from Carbon, Wyoming.

Habitat: Fossil Forest Ridge, Yellowstone National Park, bed No. 1, the lowest bed, rare; bed No. 5, rare; bed No. 6, the "Platanus bed," most abundant locality, over 75 specimens noted; bed No. 7, rare; collected by Lester F. Ward and F. H. Knowlton, August, 1887.   East end of Fossil Forest Mountain, middle bed, 775 feet above valley below; specimens rare, collected by Ward and Knowlton, August 15 and 22, 1887.   Specimen Ridge, opposite Slough Creek, rare; collected by Ward and Knowlton, August, 1887.   Hague's Yellowstone National Park collections (field No., 1960), Fossil Forest section, very abundant; collected by G. M. Wright and Walter H. Weed, September 20, 1885.   Hague's Yellowstone National Park collections (field No., 1217), Fossil Forest section, upper stratum; collected by Arnold Hague, September 24, 1884.   Hague's Yellowstone National Park collections (field No., 1219), rare; collected by Arnold Hague, September 24, 1884.   South end of Crescent Hill, 6 feet below "Platanus bed;" collected by F. H. Knowlton, August 9, 1888.

PLATANUS MONTANA n. sp.

PL. XCVI, figs. 2, 3.

Leaves membranaceous, somewhat roughened, rounded-oblong in shape, decurrent on the petiole, rounded above or acuminate, possibly slightly 3-pointed; margin simply undulate toothed; nervation obscurely palmate; petiole stout; midrib thick, straight; secondaries several (about 5) pairs, the lowest some distance above the base of the blade, emerging at an angle of about 30°, passing nearly straight to the border and ending in a small marginal tooth, with several branches on the outside approximately at right angles to the midrib and ending in marginal teeth; second pair of secondaries strong, arising at an angle of 45°, much arching upward and ending either in the margin or possibly in short lobes, with several strong forking branches on the outside, the terminations ending in the teeth; other secondaries also occasionally forked on the outside; nervilles strong, occasionally percurrent, but mainly forked or broken; finer nervation quadrangular.

This species is based on a number of more or less fragmentary leaves, the best of which are figured.   The most perfect specimen is 12 cm.

long and about 10 cm broad. It was probably, when living, at least 15 cm. long.

The marked feature of this leaf is that it is not strictly palmately nerved, having, as pointed out in the diagnosis, the 2 lower pairs of secondaries with branches on the outside which end in the marginal teeth. Otherwise it is hardly to be distinguished from *Platanus raynoldsii* Newby., as figured by Lesquereux[1] from the Denver beds of Golden, Colorado. This species was described by Newberry[2] as having the margin doubly serrate, but a number of specimens referred to it by Lesquereux have the margin undulate, dentate, or even entire. Newberry's type had 3 lobes or points in the upper portion, while certain of Lesquereux's specimens were rounded and entire above.

The smaller leaves from the lower Yellowstone described by Professor Ward under the name of *Grewiopsis populnedia*,[3] especially fig. 1 of his plate, approach the leaves under discussion. These, as he has already pointed out, are suggestive of *P. raynoldsii*. They can hardly belong to Grewiopsis.

Whether the leaves from the Yellowstone National Park should be regarded as new to science or referred to *Platanus raynoldsii* is an open question. They agree closely enough in size, shape, and marginal dentition, but differ in the nervation. It is possible that this character may be of sufficient importance to keep them distinct, and also to exclude them from the genus Platanus, but for the present at least, and until better material can be obtained, they may remain as above.

Habitat: East slope of high hill about three-fourths of a mile south from Yanceys; collected by George M. Wright, September 4, 1885.

LEGUMINOSÆ.

ACACIA MACROSPERMA n. sp.

Pl. XCVIII, fig. 8.

Legume large, more than 8 cm. long and 2.2 cm. wide, broad linear, possibly constricted, with obtuse, regularly rounded end; apparently surrounded by a wing 5 mm. broad; seeds numerous, large, oblong, 10 mm. long, 6 mm. broad.

Tert. Fl., Pl. XXVI, fig. 4; Pl. XXVII, figs. 1-3.
Later Ext. Fl., p. 629; Ill. Cret. and Tert., Pl. XVIII.
Types of the Laramie Fl., p. 80, Pl. XI, figs. 3-5.

This species appears quite unlike any species before found in America, but is not greatly unlike *A. microphylla* Heer from the Swiss Tertiary. The latter species is not quite as broad as *A. noncosperma*, and has not the end preserved. The seeds are about the same size in both.

Habitat: Fossil Forest Ridge, bed No. 7, "Castanea bed;" collected by Ward and Knowlton, August 16–20, 1887.

ACACIA LAMARENSIS n. sp.

Pl. XCVIII, fig. 6.

Legume linear, broad, more than 7 cm. long, and 1.7 cm. broad; end pointed; apparently with marginal wing 2 or 3 mm. wide; seeds oval, 10 mm. long, 8 mm. wide.

This may possibly be the same as *A. noncosperma*, but it appears to differ essentially in being narrower and in having an acuminate instead of an obtuse termination. The apparent wing and the seeds are much the same in both.

Habitat: Lamar River, between Cache and Calfee creeks; collected by Knowlton and Culver, August 21, 1888.

ACACIA WARDII n. sp.

Pl. XCVIII, fig. 7.

Legume narrow, linear, constricted, 6 cm. long, 9 mm. wide in the broadest portion and 5 mm. wide at the constricted point; point of attachment reduced to a slight extension, opposite extremity with a decided curved beak; seeds apparently present, but obscure.

This species differs markedly from the others just described, and also, so far as I know, from any heretofore found.

Habitat: Fossil Forest Ridge, bed No. 4, "Aralia bed;" collected by Ward and Knowlton, August 16–20, 1887.

LEGUMINOSITES LESQUEREUXIANA Kn.

Pl. LXXXIX, fig. 4.

Leguminosites lesquereuxiana Kn.: Bull. U. S. Geol. Surv. No. 152, 134, 1898.
Leguminosites cassinoides Lx.: Tert. Fl., p. 300, Pl. LIX, figs. 1–4.

Habitat: Northeast side of Crescent Hill opposite small pond; collected by F. H. Knowlton and G. E. Culver, August 2, 1888.

LEGUMINOSITES LAMARENSIS n. sp.

PL. LXXXIX, figs. 5, 6.

Leaflets thin, oblong-lanceolate, rounded-truncate at base, long acuminate at apex; midrib strong, perfectly straight; secondaries about 9 pairs, alternate, at an angle of 45°, slightly curving upward; remainder of nervation not retained.

This little leaflet is 6 cm. in length and 17 mm. in width. It is very regularly rounded, almost truncate at base, and apparently regularly narrowed above into an acuminate apex. The petiole, if there was one, is not preserved. The secondaries are alternate and camptodrome, and about 8 or 9 pairs.

The nearest related species is *Leguminosites lesquereuxianus*,[1] from the Green River beds of Green River, Wyoming, and also Spring Canyon, Montana. This differs in being larger, broader, and more oblong-ovate than the one under discussion. The relationship is evidently close, and perhaps more material would show closer affinity than I have recognized.

This species also resembles some of the species of Leguminosites from the Tertiary of Switzerland, as, for example, *L. proserpinæ* Heer.[2] There can be no question as to the correctness of the reference to this genus.

Habitat: East bank of Lamar River, between Cache and Calfee creeks; collected by F. H. Knowlton, August, 1888.

ANACARDIACEÆ.

RHUS MIXTA? LX.

Rhus mixta Lx., Mem. Mus. Comp. Zoöl., Vol. VI, No. 2, p. 30, Pl. IX, fig. 13.

A single small and somewhat fragmentary specimen. It resembles the smaller of the two specimens figured by Lesquereux.

Habitat: East bank of Lamar River, between Cache and Calfee creeks; collected by F. H. Knowlton, August 24, 1888.

Tert. Fl., p. 300, Pl. LIX, figs. 1-4.
Fl. Tert. Helv., Vol. III, Pl. CXXXVIII, figs. 50-55.

CELASTRACEÆ.

CELASTRUS CULVERI n. sp.

Pl. XCVII, fig. 4.

Leaves membranaceous, ovate-lanceolate, apparently rather abruptly rounded at the base, but gradually narrowed above to an obtusely acuminate apex; margin with rather remote, small, sharp, outward-pointing teeth; midrib thick below, much thinner above; secondaries about 10 pairs, alternate at an angle of 35 or more, much curved upward, camptodrome very near the margin, with branches outside entering the small, weak teeth; intermediate secondaries occasional, thin, disappearing before reaching half the distance to the margin; nervilles percurrent; finer nervation obscure.

This species is represented by 2 well-preserved leaves, both, unfortunately, representing the upper portion only. The longest specimen is 10 cm. in length, which is probably not far from its original full length. It is a little over 5 cm. broad at a point which seems to be some distance below the middle. Judging from the contour near the base, it seems probable that it was either truncate or, possibly, heart-shaped at base. The teeth of the margin are peculiar, being scattered, small, sharp, and outward pointing.

This species appears to find its nearest relative in some described from the Fort Union group along the lower Yellowstone. Thus it resembles *Celastrus curvinervis* Ward[1] in shape and nervation, but differs considerably in size and wholly in the teeth. *Celastrus ovatus* Ward[2] has somewhat the same shape, but differs considerably in nervation and in the teeth. Several of the other species described by Professor Ward[3] also resemble this in one or more particulars, but none closely enough for specific identity.

I take pleasure in naming this fine species in honor of Prof. G. E. Culver, who assisted in collecting at this place.

Habitat: Yellowstone River, one-half mile below the mouth of Elk Creek, top of bluff; collected by F. H. Knowlton and G. E. Culver, August, 1888.

Types of the Laramie Fl., p. 82, Pl. XXXVI, figs. 5, 6.
Op. cit., p. 84, Pl. XXXVI, fig. 4.
Op. cit., Pl. XXXV.

CELASTRUS INÆQUALIS n. sp.

Pl. XCVIII, fig. 5.

Leaf of firm texture, elliptical-obovate in outline, strongly unequal-sided, rounded, obtuse above, and narrowed below into an apparently winged petiole; margin strongly sinuate-dentate from above the lower third or half of the blade; midrib thin, approximately straight; secondaries 10 or 12 pairs, lower pairs opposite, others alternate, two lower pairs, and about 6 secondaries on the larger side of the blade at right angles to the midrib, those on narrow side of blade and in upper portion of other side from right angle to 45° or more, all camptodrome, arching upward abruptly near the margin and apparently joining, sending branches from the outside to the teeth and other parts of the margin; nervilles and finer nervation obsolete.

This species, as exemplified by the specimen figured, is very peculiar. It is 7 cm. long and a little more than 4.5 cm. wide. It is abruptly obtuse at apex and appears to be expanded at base into a winged petiole.

The margin in the lower portion is quite entire, while above it is strongly sinuate-toothed. The nervation is markedly peculiar. The lower secondaries are at right angles with the midrib, as are several more on the broader margin, while those on the narrow side of the blade all arise at a less angle and curve abruptly near the margin. Those in the extreme tip of the blade curve very much after the manner of Cornus. All send branches from the outside to the teeth. The finer nervation is not preserved.

This species is wholly unlike any with which I am familiar. It possesses the character described by Professor Ward as especially characteristic of the American fossil forms of this genus, namely, the arching of the secondary nerves near the extremities, with short subsidiary nerves to the marginal teeth. Its unequal-sidedness, winged petiole, and sinuate teeth above the middle of the blade seem to still further characterize it.

Habitat. Yellowstone River, one-half mile below the mouth of Elk Creek, base of bluff; collected by F. H. Knowlton, August, 1888.

Types in the Laramie U. p. 58.

CELASTRUS ELLIPTICUS n. sp.

PL. XCVII, fig. 3.

Leaf of firm texture, nearly regularly elliptical in outline, abruptly rounded above to an obtuse apex and below to an almost truncately rounded base which is slightly decurrent along a short petiole; margin irregularly sinuate-toothed from a short distance above the base; midrib rather thick, passing straight to the apex; secondaries about 15 pairs, alternate or sub-opposite, at an even angle of about 35°, straight; distal termination of secondaries unknown; nervilles and finer nervation obsolete.

This perfect leaf is 7 cm. long and 4.5 cm. broad. It is very slightly unequal-sided, the difference being, however, hardly 3 mm. It is very regularly elliptical, with a sinuate-dentate margin, which begins about one-fourth the length of the leaf from the base, the lower portion being entire. The nervation is very regular, consisting of about 15 pairs of secondaries, which emerge at an angle of about 35° and run straight toward the margin, but the manner of the termination at the margin can not be made out, from lack of preservation. It is probable that they arch abruptly near the margin and send secondary nervilles to the teeth. None of the nervilles or finer nervation can be made out.

It is possible that this species is closely related to *C. inaequalis*, just described, as they come from the same beds, but it seems hardly probable. This latter species, as already pointed out, is very unequal-sided, with large sinuate teeth, and peculiar arrangement of secondaries. *C. ellipticus*, on the other hand, is almost regular in shape, has twice as many and smaller sinuate teeth, and regular secondaries.

This species does not approach closely to any described species of Celastrus with which I am familiar.

Habitat: Yellowstone River, one-half mile below the mouth of Elk Creek, at base of bluff; collected by F. H. Knowlton, August, 1888.

ELAEODENDRON POLYMORPHUM Ward.

PL. XCVII, fig. 1.

Elaeodendron polymorphum Ward: Types of the Laramie Fl., p. 84, Pl. XXXVIII, figs. 1–7.

The fine specimen figured is referred with some doubt to this species. It has much the same shape, serration, and type of nervation as *E. poly-*

morphum, but differs in being much larger and in having more numerous and closer secondaries. It is undoubtedly close to this species, and rather than make it a new species I have referred it to this.

Habitat: Yancey Fossil Forest, near the standing trunks; collected by F. H. Knowlton, August, 1888.

ACERACEÆ.

ACER VIVARIUM n. sp.

Pl. XCVIII, fig. 4.

Leaf membranaceous, palmately 3-lobed, narrowed below to a wedge-shaped base, sinuses rounded, middle lobe lanceolate-acuminate, as long or longer than the body of the blade below the sinuses; lateral lobes at an acute angle (points not preserved); margins remotely toothed with small, sharp, upward-pointing teeth; midrib, or central rib, strong, straight, slightly stronger than the lateral ribs, which arise from the midrib just above the base of the blade at an angle of about 70° and pass up straight to the points of the lateral lobes or curve slightly outward; lateral ribs with several pairs of secondary branches, those on the outside beginning just above the base of the blade and passing straight or with a slight upward curve to or entering the teeth; secondaries on the upper or inside, beginning below the sinus, which the lowest one enters, the others probably entering the teeth; middle lobe with about 6 pairs of alternate secondaries arising at an angle of 70° or 75°, and passing up nearly straight, to end in the teeth or fork just below the teeth, one branch entering and the other going upward near the margin to the one above; nervilles numerous, mainly percurrent; finer nervation beautifully preserved, forming quadrangular areolæ.

The example figured is the only one observed of this species. It is about 10 cm. long and 6 cm. broad. The central lobe is about 5 cm. long and a little more than 2 cm. wide. The lateral lobes appear to have been about 2 cm. wide and of an unknown length.

This leaf belongs to the *Acer trilobatum* group, so many forms of which were described by Heer from the Swiss Tertiary. In shape it is most like *A. trilobatum productum*,[1] but differs in having only very small, sharp teeth. It is

---

[1] Foss. Helv., Vol. III, Pl. CXV, figs. 6–12.

also somewhat like *A. trilobatum tricuspidatum* Heer, as figured by Professor Ward from the Fort Union' group, differing in being much more wedge-shaped at base, and in the angle of the lateral ribs and secondaries.

Habitat: Fossil Forest Ridge, bed No. 3, "Magnolia bed;" collected by Ward and Knowlton, August, 1887.

ACER, fruit of.

Pl. XCVIII, fig. 5.

The collection contains several of these fruits, the best of which is figured. While they are very definite and clearly belong to Acer, they are not usually regarded as being sufficiently distinctive for specific reference. A number have been figured and named also, but I have preferred not to name these. They may possibly belong to the preceding species, but of this there is no proof.

Habitat: Crescent Hill above Yanceys; collected by W. H. Weed, September 28, 1885.

SAPINDACEÆ.

SAPINDUS AFFINIS Newby.

Pl. CII, figs. 1–3.

Sapindus affinis Newby.; Later Extinct Floras, p. 51; Ill. Cret. and Tert. Fl., Pl. XXV, fig. 1.

The material upon which this determination is based is ample, as it consists of fully 100 specimens, all more or less well preserved. These specimens differ so much in size that it was at first thought that at least 2 species must be represented, but after a careful study it has been found impossible to draw any satisfactory line between them. There are all gradations of size from the little slender leaflets, hardly 4 cm. long, to the large ones, fully 10 cm. long.

In the only published figures of this species by Newberry the nerva-tion is not shown, but the National Museum contains the original New-berry material, and on studying this it is found that the nervation agrees perfectly with the specimens from the Yellowstone National Park. It may

Types of the Laramie Fl. Pl. XXIX, fig. 5.

also be noted that Newberry's material does not show the leaflets as petioled, but in the description of *S. affinis* he says, "leaflets . . . . sessile or short-petioled." Many of the detached leaflets in his material, named in his handwriting, are distinctly short-petioled, in which particular the Park specimens agree. Some have, it is true, very short petioles, yet all seem to have them.

In only two cases have leaflets been found in this collection attached to the rachis, and these have both been figured.

Habitat: Yellowstone River, one-half mile below the mouth of Elk Creek; found throughout the section, and most abundant at the bottom; collected by F. H. Knowlton, August, 1888.

? SAPINDUS ALATUS Ward.

Sapindus alatus Ward: Types of the Laramie Fl., p. 68, Pl. XXXI, figs. 3, 4.

This specimen is the only one that I venture to refer to this species. It was found in the same beds with the abundant *S. affinis* Newby., and possibly may be an abnormal or unusual leaflet of that species. It is, however, more regular, and has the thick or winged petiole of *S. alatus*.

Habitat: Yellowstone River, one-half mile below the mouth of Elk Creek, Yellowstone National Park; top of bluff, 500 feet above the river; collected by F. H. Knowlton, August, 1888.

SAPINDUS GRANDIFOLIOLUS Ward.

Pl. XCIV, figs. 1, 2; Pl. CII, fig. 4.

Sapindus grandifoliolus Ward: Types of the Laramie Fl., p. 67, Pl. XXX, figs. 3–5; Pl. XXXI, figs. 1–2.

Several small doubtful leaves are referred to this species. One in particular has some resemblance to leaves of *Juglans rugosa* Lx., but seems to be closer to the *Sapindus grandifoliolus* of Ward.

Habitat: Fossil Forest Ridge, bed No. 6, "Platanus bed;" collected by Ward and Knowlton, August, 1887. Also found on the south side of Stinkingwater Valley, on a high bluff east of mouth of Crag Creek, collected by F. P. King for Arnold Hague, September 1, 1897.

SAPINDUS GRANDIFOLIOLOIDES n. sp.

Pl. C, fig. 2.

Leaflets large, of firm texture, ovate-lanceolate, unequal-sided, rounded at base to a well-marked winged petiole, apex acute, slightly falcate; midrib of moderate strength, straight; secondaries about 7 or 8 pairs, strongly alternate emerging at a low angle and soon curving up and passing along near the border; the secondaries on the narrower side of the leaflet emerge at a greater angle (30° to 45°) than on the opposite side; finer nervation not preserved.

The specimen figured is absolutely perfect so far as outline and principal nervation go. It is just 10 cm. long and 3.7 cm. wide in the broadest portion, which is a little below the middle. The margin is slightly undulate, almost toothed in one part.

This species so closely resembles *Sapindus grandifoliolus* Ward[1] from the Fort Union group, that I have named it *grandifolioloides*. It differs, however, from the latter in being more markedly inequilateral and in having a winged petiole. It also has fewer secondaries than *S. grandifoliolus*.

This species is also related to *S. obtusifolius* Lx.,[2] found in the Fort Union group. From this it differs in having half the number of secondaries and a winged petiole, otherwise being much the same.

Professor Ward's *S. alatus* from the same place as *S. grandifoliolus* has a winged petiole, but differs in being much smaller and in having a broken, loose nervation. It is possible, however, that if more material could be obtained it could be referred to one or the other of these evidently related forms.

Habitat: Northeast side of hill above Lost Creek, bed No. 1, collected by F. H. Knowlton, August 8, 1888.

SAPINDUS WARDII n. sp.

Pl. XCVIII, figs. 1, 2; Pl. XCIX, fig. 5.

Leaflet coriaceous, broadly lanceolate, rounded wedge-shaped at base, with long acuminate falcate apex; margin perfectly entire; midrib thick,

---

Types of the Laramie Fl., p. 67, Pl. XXX, figs. 3-5, 1887.
Cret. and Tert. Fl., p. 253, Pl. XLVIII, figs. 5-7, 1888.
Op. cit., p. 68, Pl. XXXI, fig. 3-4, 1887.

passing through the middle of the blade; secondaries about 8 pairs, alternate, strongly camptodrome, forming broad loops at a marked distance from the margin, occasionally with a series of smaller loops outside; intermediate secondaries occasional; nervilles few, percurrent; finer nervation forming large quadrangular meshes.

The specimens figured are the only ones observed of this species. The best preserved is a little over 10 cm. in length and 4 cm. broad. It is marked by the long, slender, and falcate apex, and the peculiar looped secondaries, which are joined far inside the margin. One side of the basal portion of the leaflet is missing, but from the direction of the secondaries it is probable that it was somewhat unequal-sided.

Fig. 1, Pl. XCVIII, which lacks both base and apex, must have been at least 13 cm. in length, and was probably longer. The other (fig. 2, Pl. XCVIII) was about the size of the best-preserved example.

These leaflets very closely resemble *Fraxinus affinis* Newby., from Bridge Creek, Oregon. This has the same type of nervation, but is much smaller, very slightly unequal-sided, and with more numerous and more regular looped secondaries. The finer nervation is identical in each.

There is some doubt as to the correctness of the reference of Newberry's leaf to the genus Fraxinus. This much resembles the genus Sapindus and may possibly belong to it. *Sapindus grandifoliolus* Ward, from the Fort Union group, for example, has much resemblance in general character to these leaves. It would seem that they should all be in the same genus. However, the leaflet under consideration is undoubtedly closely allied to what Ward has called Sapindus, and for the present they may remain there.

I have named this characteristic species in honor of Prof. Lester F. Ward, who was present when it was collected.

Habitat: Fossil Forest Ridge, bed No. 5; collected by Lester F. Ward and F. H. Knowlton, August 16–19, 1887. Yellowstone River, one-half mile below mouth of Elk Creek; collected by F. H. Knowlton, August, 1888.

U. S. Nat. Mus., Vol. V, 1882, p. 510, Plates incl. Pl. XLIX, fig. 3.
Types of the Laramie Fl., p. 67, Pl. XXX, figs. 4, 5.

RHAMNACEÆ.

RHAMNUS RECTINERVIS Heer.

Rhamnus rectinervis Heer: Fl. Tert. Helv., Vol. III, p. 80, Pl. CXXV, figs. 2, 6. Lesquereux: Tert. Fl., p. 279, Pl. LIII, figs. 12–15.

This species was first detected in the Park by Lesquereux,[1] and the present collection contains a number of specimens that may be similarly referred. They are all entire, in which respect they resemble fig. 14 of Lesquereux's plate (loc. cit.).

Habitat: Fossil Forest Ridge, bed No. **3**, "Magnolia bed;" bed No. **7**, "Castanea bed," collected by Ward and Knowlton, August, 1887.

PALIURUS COLOMBI Heer.

Pl. CI, fig. 7.

Paliurus colombi Heer. Lesquereux: Tert. Fl., p. 273, Pl. L, figs. 14–17 (1878).

The little specimen figured appears to be the only one obtained in the area under discussion. It is of the same size and shape as many of the figures of this species, but seems to differ slightly in having 2 strong secondaries on one side of the midrib. It, however, approaches certain of the figures given by Heer[2] of arctic examples of this species.

Habitat: Head of Tower Creek; collected by W. H. Weed, September 25, 1885.

ZIZYPHUS SERRULATA Ward.

Pl. CI, figs. 4, 5.

Zizyphus serrulata Ward: Types of the Laramie Fl., p. 75, Pl. XXXIII, figs. 3, 4 (1887).

The two figured examples agree very closely with the figures given by Professor Ward. They both have teeth well marked, and thus agree with fig. 4 (loc. cit.). They are not quite so well preserved as the types, and do not show the finer nervation, but there can be no doubt as to their identity.

Habitat: Yellowstone River, one-half mile below the mouth of Elk Creek, top of bluff; collected by F. H. Knowlton, August, 1888.

Hayden's Ann. Rept., 1878, Pt. II, p. 19.
Fl. Foss. Arct., Vol. I, Pl. XIX, fig. 5.

VITACE.E.

Cissus HAGUEI n. sp.

Pl. Cl, fig. 2.

Leaf membranaceous, quadrangular-ovate, truncate or possibly slightly heart-shaped at base and acuminate at apex, lateral lobes short, obtuse; margin toothed, the teeth low, obtuse or somewhat acute; nervation palmate, midrib thin, perfectly straight, lateral ribs of same strength as midrib, arising at an angle of 45°, passing directly to and terminating in the obtuse lateral lobes; ribs with 4 or 5 branches on the outside, which terminate in marginal teeth, secondaries about 4 pairs, alternate, at same angle as the ribs and terminating in the teeth; nervilles thin, sparse, percurrent or often broken.

This fine leaf is 8.5 cm. long, 5.2 cm. broad between the lobes and 6.5 cm. broad in the widest part, which is only a short distance above the base. In outline it is what may be called quadrangular-ovate—that is, between ovate and square. It is palmately 3-ribbed, the lateral ribs being at an angle of about 45° and of the same strength as the midrib. They pass straight to and terminate in the short lateral lobes, and have 4 or 5 outside branches which also terminate in the marginal teeth.

The relation of this species is undoubtedly with *Cissus parrottiæfolia* Lx., from the Green River group. This latter species differs in being relatively longer, without especially marked lateral lobes, with larger, more obtuse teeth, and unforked outer branches of the lateral ribs. There are also more secondaries in the upper part of the leaf. These, however, are but slight differences, and are possibly only such as might be expected in individual variation. But as only one example has been found in the Yellowstone National Park, there is no means of knowing what may be allowed for individual variation, so I have preferred to keep them separate.

I take pleasure in naming this species in honor of Mr. Arnold Hague, who collected it.

Habitat: Fossil Forest Ridge, middle stratum; collected by Arnold Hague, September 24, 1884.

STERCULIACEÆ.

PTEROSPERMITES HAGUEI n. sp.

Pl. XCIX, fig. 4.

Leaf of firm texture, broadly oblong in outline, slightly inequilateral, truncate at base, obtusely pointed at apex; margin, except at base, irregularly serrate, the teeth small, sharp, upward pointing; midrib strong, flexuose; secondaries 6 pairs, alternate, at an angle of about 45°, flexuose, craspedodrome, or subcamptodrome, with branches outside entering the teeth; lower pair of secondaries forming a series of broad loops; nervilles strong, mainly broken; finer nervation not preserved.

The figured specimen of this species is 11 cm. long and nearly 7 cm. broad. As stated, it is quite regularly broad-oblong in shape, with sparsely toothed margins except near the base. The lower pair of secondaries form a series of broad loops, while the upper ones are mainly craspedodrome.

This species is evidently quite closely related to *P. minor* Ward[1] from the Fort Union group near the mouth of the Yellowstone. Fig. 2 (loc. cit.) is especially like this species, but differs in being hardly half the size and in being markedly heart-shaped at the base. It perhaps should be referred to this species except for the fact that the other 2 leaves included by Professor Ward are so very different that they can hardly be allied to the one under discussion.

Habitat: Fossil Forest Ridge; collected by Arnold Hague (No. 1249).

CREDNERIACEÆ.

CREDNERIA? PACHYPHYLLA n. sp.

Pl. CI, fig. 6.

Leaf large, thick, round-ovate, rounded and truncate or very slightly heart-shaped at base, abruptly acuminate at apex; margin apparently coarsely sinuate-toothed; petiole long (4.5 cm.), thick (4 mm.); midrib thick, passing to the apex; secondaries 6 or 7 pairs, the 3 lower pairs (of which the very lowest is slender and near the margin) opposite and arising from almost the same point at the base of the blade, others alternate, all

[1] Types of the Laramie Fl., p. 95, Pl. XLII, fig. 1-3.

seemingly craspedodrome, occasionally branching near the margin; finer nervation not preserved.

The specimen figured is the only one that has been found. It is 14 cm. long without the petiole, which is 1.5 cm. in length. Both sides of the leaf are destroyed, but it was probably about 10 cm. wide.

Habitat: Yellowstone River, one-half mile below mouth of Elk Creek, bluff about 40 feet above the river; collected by F. H. Knowlton, August 27, 1888.

TILIACEÆ.

TILIA POPULIFOLIA LX.

Tilia populifolia Lx.: Cret. and Tert. Fl., p. 179, Pl. XXXIV, figs. 8, 9.

A single large, fairly well preserved specimen is all that has been found of this species. It is referred with very little doubt to Lesquereux's species, which was before known only from Florissant, Colorado. It is a little less heart-shaped at base than fig. 8 (loc. cit.) of Lesquereux's plate, but in the discussion of this species Lesquereux describes it as "round or subcordate at base." The teeth are of precisely the same character, being only slightly smaller. The thick petiole and fine palmate nervation are identical, as is the other secondary nervation.

Habitat: Yellowstone River, one-half mile below the mouth of Elk Creek, top of bluff; collected by F. H. Knowlton, August 27, 1888.

GREWIOPSIS ? ALDERSONI n. sp.

Leaves of firm texture, broadly obovate, truncate or slightly heart-shaped at base, obtusely acuminate above; margin entire at base, slightly undulate-toothed above; midrib thick, straight, nervation pinnate; secondaries about 6 pairs, alternate, at an angle of 45°, camptodrome; lowest pair subopposite, arising some distance above the base of the blade, with 3 or 4 tertiary branches from the outside which are camptodrome and arch well inside the margin; upper secondaries occasionally forked near the margin; nervilles strong, percurrent.

I refer several specimens to this somewhat doubtful species. Neither of them are perfect, but as far as can be made out the average length appears to have been about 9 cm. and the width about 6 cm.

It is doubtful if these leaves belong to the genus Grewiopsis, but at present I am unable to suggest a closer affinity. They have the same size, shape, and approximately the same nervation as *G. platanifolia* Ward,[1] from the Fort Union group, differing, however, in not having a toothed margin. Professor Ward writes that the specimen upon which his species was founded is quite obscure, and it is possible that they may really be nearer alike than appears from the drawings. Additional material is needed to fix their status.

I have ventured to call this a new species, and have named it in honor of Mr. E. C. Alderson, who accompanied the expedition on which it was obtained and assisted in making the collections.

Habitat: Specimen Ridge, opposite mouth of Slough Creek and near head of Crystal Creek; collected by Ward and Knowlton, August, 1887.

ARALIACEÆ.

ARALIA WRIGHTII n. sp.

Pl. CI, fig. 1.

Leaf firm, coriaceous, narrow in general outline, palmately 3 (possibly 5) lobed; central lobe largest, long, ovate-lanceolate, slender-pointed; lateral lobes slender-lanceolate, half the length of the central lobe; all margins perfectly entire; basal portion of leaf unknown; primary nervation palmate; middle lobe with a pair of opposite nerves nearly at right angles to the midrib, which pass to the sinus, those above with about 10 pairs of alternate camptodrome secondaries, which are much curved upward and arched along near the margin; intermediate secondaries occasional; lateral lobes with a strong midrib and about 8 pairs of alternate or subopposite much arched camptodrome secondaries; finer nervation consisting of very fine quadrangular areolation.

This very peculiar species is unfortunately represented by only the fragmentary leaf figured. The basal portion is entirely destroyed and it is therefore impossible to determine whether there were 5 or only 3 lobes. There is some evidence in favor of its having been 5-lobed. The sinuses separating the lobes are somewhat rounded. The central lobe is very much the larger. From the sinus it has 6 cm. preserved and must have been 8 cm. or more in length when entire. In the broadest part,

[1] Types of the Laramie Fl., p. 89, Pl. XI, fig. 1.

which is about one-fourth of its length from the sinus, it is 2.5 cm. broad. The lateral lobes are about 5.5 cm. long, not enlarged upward. At base they are 4 cm. broad, from which point they taper gradually to a slender acuminate apex. The nervation has been described in the diagnosis, and may also be clearly made out from the excellent figure.

It is hardly possible to compare this species with described forms, from the fact that it is so fragmentary that the perfect form can not be made out. The characters of the larger middle lobe and the very much smaller lateral lobes seem to be so marked that it is strongly separated from any described species. *Aralia angustiloba* Lx.[1] from the Chalk Bluffs of California, perhaps is closest to this species, yet it differs markedly. It will be necessary to wait for additional material before its exact character can be made out.

I have named this species in honor of Mr. George M. Wright, one of the collectors of this and many other valuable specimens in the Yellowstone National Park.

Habitat: Fossil Forest (No. 22c of section), collected by Wright and Weed, September 20, 1885.

ARALIA NOTATA LX.

Pl. C, fig. 4.

Aralia notata Lx.: Tert. Fl., p. 237, Pl. XXXIX, figs. 2-4. Ward: Types of the Laramie Fl., p. 60, Pl. XXVII, fig. 4.
Platanus dubia Lx.: Hayden's Ann. Rept. 1873-1874, p. 405.

The collections contain about 50 specimens that evidently belong to this species. None of them are absolutely perfect, yet the general character can be made out. They come from three localities, one of which, the Yellowstone below Elk Creek, was given as a type locality by Lesquereux.[2]

There appears to have been a tendency among later writers to regard this as the same as Newberry's *Platanus nobilis* from the Fort Union group, which indeed it much resembles. They were both very large species, not often preserved entire, but they seem to differ essentially. On this point Lesquereux says: "This species (*A. notata*) seems very closely allied to

Mem. Mus. Comp. Zoöl. Vol. VI, No. 2, Pl. V, figs. 4, 5.
Tert. Fl., p. 237.
Later Extinct Flora, p. 67.

Platanus nobilis Newby.; I should not hesitate to consider it as identical, but for the character of the lateral nerves, which are described by the author as straight, and terminating in the teeth of the margin. In this species the borders are entire and the lateral nerves camptodrome. The difference may be merely casual, for one of the specimens from Troublesome Creek has the close secondary veins camptodrome along the borders of the inner side of the lobes, while on the outer side the borders are obscurely cut by a few small teeth, into which the veins enter as craspedodrome. Other specimens, thus of Elk and Yellow creeks,[1] have the characters of *P. nobilis*."

It would thus appear that Lesquereux himself inclined to regard the Park specimens as being referable to *Platanus nobilis*, but in the 50 or more specimens that I have studied from this place I have not found one showing the teeth and craspedodrome nervation of *P. nobilis*. They all have the distinctly camptodrome nerves, as shown in Lesquereux's figures. I have therefore decided to keep them under *Aralia*.

The further question of the correctness of this generic reference, or rather of the relation of this *Aralia notata* to the genus Platanus, will not now be taken up. Janko has said[2] that *Platanus nobilis* "non est Platanus," while on the other hand Professor Ward has suggested[3] that several of the so-called species of Aralia may have to be united into a group, under the name of Protoplatanus, representing the ancestors of Platanus. A small specimen of this species, obtained by Prof. J. P. Iddings from a gulch northeast of the peak west of Dunraven, is exceptionally well preserved, at least as regards the finer nervation. This is very regularly square, being only about 0.25 mm. in size. The leaf appears to have been rather thick, possibly coriaceous.

No other specimen that I have seen has this finer nervation so well preserved.

Habitat: Fossil Forest Ridge, Yellowstone National Park, bed No. 7, "Castanea bed", about 25 specimens; collected by Lester F. Ward and F. H. Knowlton, August 16–20, 1887. Southeast end of hill above (north) Lost Creek, bed No. 4, 2 leaves; collected by F. H. Knowlton, August 8, 1888. Yellowstone River, one-half mile below mouth of Elk

[1] Probably Elk Creek on Yellowstone River. F. H. K.
[2] Abstammung d. Platanen. Englers bot. Jahrb., Vol. XI, 1889, p. 458.
[3] Types of the Laramie Fl., p. 83.

Creek, top of bluff; collected by F. H. Knowlton, August 27, 1888. Andesitic breccia, near gulch northwest of peak west of Dunraven; collected by J. P. Iddings, September 12, 1885. Also found on Overlook Mountain, in breccia, at an altitude of 10,070 feet, collected by Arnold Hague, August 6, 1897. Southern spur of Chaos Mountain, at an altitude of 10,100 feet; collected by Arnold Hague, August 11, 1897. South side of Stinkingwater Valley, on high bluff east of mouth of Crag Creek; collected by Arnold Hague, September 4, 1897.

ARALIA SERRULATA n. sp.

Pl. CI, fig. 5.

Leaf apparently subcoriaceous, palmately 3-lobed, middle lobe longest, ovate, obtuse; lateral lobes short, pointing upward; borders sharply serrulate, with small, sharp, upward-pointing teeth, secondaries numerous, close, alternate, at an angle of 25° to 40°, curving upward and entering the teeth, or sometimes camptodrome with outside branches to the teeth, usually 1 tooth between the 2 entered by two contiguous secondaries, which is supplied with a branch from the middle of a percurrent nerville, which crosses just below it; nervilles numerous, mainly percurrent and approximately at right angles to the secondaries; finer nervation quadrangular.

This fine and apparently characteristic species depends upon the single example figured. It lacks the entire lower portion of the leaf, but 2 lobes are entirely preserved, and a large portion of the other. The central lobe is 4.5 cm. long to the sinus, and the lateral one about 1 cm. higher than the sinus. The distance between the lateral lobes is 8.5 cm.

This species has exactly the same size and shape as many of the 3-lobed specimens of *Aralia notata* Lx.,[1] found in the same beds.

The main difference is in the sharply serrate margins, the teeth extending even down to and through the sinus, and in the secondaries or branches from them entering the teeth. Occasionally, as indicated under the diagnosis, some of the secondaries are camptodrome, as all are in *A. notata*, with outside branches passing to the teeth. These species are evidently closely related and may possibly be the same, although probably not, for in 100 specimens of *A. notata* not one was found that possessed these teeth.

As pointed out under the discussion of *Aralia notata* (see ante, p. 745),

Cf. Lesquereux, Tert. Fl., Pl. XXXIX, figs. 2, 3. Ward, Types of Laramie Fl., Pl. XXVII, fig. 1.

there was some tendency to refer it to *Platanus nobilis* Newby., which is sometimes slightly toothed. The teeth of the species under discussion are distinctly aralioid, and not at all like those of *P. nobilis*.

Aralia serrulata is distantly related to *A. digitata* Ward,[1] from the Fort Union beds. This latter species is 3-lobed, or, more often, 5-lobed, with the lobes enlarged upward, and serrate with shallow teeth only near the apex. *A. macrophylla* Newby.,[2] from the Green River group of Wyoming, has the lobes serrate, but the teeth are large, coarse, and often scattered, and, moreover, the leaf is twice the size of this and always 5-lobed.

A number of species of Aralia have been described from California, but none of them agree with *A. serrulata*.

Habitat: Yellowstone River, one-half mile below the mouth of Elk Creek, top of bluff; collected by F. H. Knowlton, August, 1888.

ARALIA WHITNEYI LX.

PL. XCIX, fig. 5.

Aralia whitneyi Lx., Foss. Pl. Aurif. Gravels. Mem. Mus. Comp. Zoöl., Vol. VI., No. 2, 1878, p. 20, Pl. V, fig. 1. Hayden's Ann. Rept. 1878, Pt. II, p. 49.

This fine species was described by Lesquereux from the Auriferous gravels of Chalk Bluff, Nevada County, California, and was also recognized by him in material collected by Mr. W. H. Holmes on Fossil Forest Ridge[2] in 1878. The specimens here referred to this species come from probably the same locality as that which afforded Holmes material. They are, with one exception, larger leaves than described in the type. None of the specimens are perfect, and hence it is difficult to determine the exact size, but they must have been 15 to 20 cm. long and probably broader.

The small specimen mentioned is referred with some hesitation to this species. It is only about 9 cm. broad and 7 cm. long, but otherwise hardly differs.

Habitat: Fossil Forest Ridge, Yellowstone National Park, bed No. 4, "Aralia bed," small leaf only; bed No. 7, "Platanus bed;" Specimen Ridge, Fossil Forest, opposite Slough Creek, and near head of Crystal Creek, "Platanus bed," several large fine leaves.

Types of the Laramie Fl. p. 67, Pl. XXVIII, fig. 1.

Proc. U. S. Nat. Mus. 1887, p. 535, Plates med., Pl. LXVII, fig. 1; Pl. LXVIII, fig. 1.

Cf. Hayden's Ann. Rept., 1878, Pt. II, p. 79.

ARALIA sp.

This fragment is the only one of this type observed, and is too poor to admit of satisfactory identification or characterization if it be new. It consists of a portion of what appears to be the central lobe and 2 lateral lobes of a 3-lobed form. The sinuses are rounded and the middle lobe is enlarged above, with the margins entire. A secondary nerve passes up to the sinuses, and the lobe has about 5 or 6 pairs of alternate much arched camptodrome secondaries. It is quite unlike any other form observed, so far as can be made out.

Habitat: Hague's Yellowstone Park collection, Fossil Forest section, No. 22; collected by Wright and Weed, September 20, 1885 (field No. 1959).

CORNACEÆ.

Cornus Newberryi Hollick.

Pl. CIII, fig. 6.

Cornus Newberryi Hollick, in Knowlton: Bull. U. S. Geol. Surv., No. 152, p. 77, 1898.
Cornus acuminata Newby: Later Extinct Floras, etc., Ann. Lyc. Nat. Hist. New York, Vol. IX, 1868, p. 71; Ill. Cret. and Tert. Pl., Pl. XX, figs. 2-4; Plates (med.), Pl. XXXVII, figs. 2-4.

Represented by a number of well-preserved leaves, agreeing well with Newberry's figures and description.

Habitat: Yellowstone River, one-half mile and also 1 mile below mouth of Elk Creek, at top of bluff; collected by F. H. Knowlton. Also found on south side of Stinkingwater Valley on high bluff east of the mouth of Crag Creek; collected by Arnold Hague, September 4, 1897.

Cornus Wrightii n. sp.

Pl. CIII, figs. 4, 5.

Leaves of firm texture, elliptical-lanceolate, narrowed below and apparently slightly decurrent, rather obtuse at apex; margin perfectly entire; midrib rather thin, slightly flexuose, secondaries 4 or 5 pairs, lower pair opposite, others alternate, at various angles, curving along the margin and in the upper part, turning by a broad bow to the apex; nervilles few, approximately at right angles to the midrib; finer nervation not preserved.

Several specimens of this interesting species are known. The most perfect one is figured, and is 7 cm. in length, and yet lacks a small portion of both base and apex. It is a little more than 2.5 cm. broad. It is quite regularly elliptical-lanceolate in shape, with a rounded, rather obtuse apex and a more narrowed base. The secondaries appear to be uniformly of 4 pairs, those in the upper portion of the leaf arching around and entering the point.

This species has some resemblance to the preceding species, which differs, however, in being much larger and in having an acuminate apex and numerous (8 or 9) secondaries. They can not be identical.

It differs from *Cornus ovalis* Ln.,[1] from Table Mountain, California, in shape and nervation, this species being oval, with obtuse base and apex.

Among living species this has considerable affinity with *C. paniculata* l'Her., especially with certain of the narrow-leaved forms.

I have named this species in honor of Mr. George M. Wright, one of the collectors.

Habitat: Fossil Forest section, Hague's Yellowstone Park collection, No. 22c of section; collected by Wright and Weed, September 20, 1885.

ERICACEÆ.

ARCTOSTAPHYLOS ELLIPTICA n. sp.

PL. XCVII, fig. 2.

Leaf very thick, leathery; elliptical in shape, obtuse above, slightly wedge-shaped at base; midrib thick, slightly flexuous; secondaries about 5 pairs, alternate, lower ones short, at a low angle, upper ones at an angle of about 45°, soon curving upward and arching about near the margin to join the one next above; nervilles strong, percurrent; finer nervation obsolete.

This fine little leaf is almost perfect. It is 4.5 cm. in length and 18 mm. in width. The petiole is about 3 mm. long and is very thick, as is the flexuous midrib. The secondaries are also strong, the upper ones arching and joining in the upper part of the leaf.

This leaf is very thick, showing that it was of firm, leathery texture. It is evidently related to the bearberry (*Arctostaphylos uva-ursi*) in shape, texture of the leaf, and nervation. It differs in being almost twice the size of the living form and in having coarse nervation. It has also a short petiole.

Habitat: Yellowstone River, one-half mile below mouth of Elk Creek; collected by F. H. Knowlton.

EBENACEÆ.

DIOSPYROS BRACHYSEPALA Al. Br.

Diospyros brachysepala Al. Br. Ward: Types of the Laramie Fl. p. 104, Pl. XLIX, figs. 1, 2.

A finely preserved leaf, almost identically the same as Ward's fig. 2, except that the secondaries are a little closer together.

Habitat: Fossil Forest Ridge, bed No. 1, "Aralia bed;" collected by Ward and Knowlton, August, 1887.

DIOSPYROS LAMARENSIS n. sp.

Pl. XCV, figs. 5, 6; Pl. XCVI, fig. 1.

Leaf membranaceous, regularly elliptical or ovate-elliptical, equally rounded at base and apex, or slightly broader at base; petiole not preserved, apparently with a slight wing; midrib thin, straight; secondaries 7 or 8 pairs, alternate, thin, camptodrome, arising at an angle of 45° or 50°, passing straight toward the borders, near which they arch and join by loops to the secondary next above; intermediate secondaries occasional, thin, usually joining the secondary next below; finer nervation consisting of numerous irregular nervilles, producing irregularly quadrangular areolæ.

This species is about 5 cm. long and a little more than 3 cm. wide, and is quite regularly elliptical in shape. As stated, the petiole is not preserved, but judging from the base of the blade it seems probable that it was slightly winged. The lower pair of secondaries arise from the very base of the blade and are very thin; the others are all alternate and camptodrome. One of the other leaves figured is approximately of the same shape, but has slightly more indication of having had a winged petiole. It is rounded at base and has a loose nervation, as in the other.

This species is closely related and possibly identical with *Diospyros ovata* Lx.[3] from Florissant, Colorado. This latter species differs in being

Tert. Fl. p. 252, Pl. XL, fig. 11; Cret. and Tert. Fl. p. 175, Pl. XXXIV, fig. 7.

more or less distinctly wedge-shaped at base, and is obovate rather than elliptical in shape. The nervation is similar in both.

D. lamarensis is also like certain leaves of *D. brachysepala* Al. Br., from Florissant and the Fort Union group of Montana.[1] The leaves from the Fort Union group are rather larger, and have the secondaries at a different angle and are without the peculiar finer nervation. It seems best, however, to keep them distinct, at least for the present.

Habitat: Lamar River, between Cache and Calfee creeks; collected by F. H. Knowlton, August, 1888. Fossil Forest; collected by Arnold Hague, September 24, 1884.

DIOSPYROS HAGUEI n. sp.

Pl. C, fig. 3.

Leaf coriaceous, elliptical, entire, obtuse at apex and base; petiole thick; midrib thick, flexuose; secondaries about 6 pairs, alternate, very irregular, the pair at the base of the blade thin, vanishing near the margin, next pair strongest, passing to the upper part, camptodrome, branching on the outside and forming broad loops well inside the margin; upper secondaries smaller, camptodrome, forming broad loops; nervilles sparse, strong, percurrent; finer nervation obsolete.

This species rests on the fine, nearly perfect leaf figured. It is 7 cm. long, including the petiole, which is 14 mm. long and 2 mm. thick. The blade is nearly regularly elliptical in shape and 3.5 cm. broad. The nervation is peculiar, as may be drawn from the description and figure. All of the secondaries except the lower pair are camptodrome, forming by union with the one next above a series of broad loops some distance inside the margin.

This type of nervation is peculiar and is clearly that of Diospyros. It approaches quite closely to certain small-leaved forms of *D. virginiana* L. Among fossil forms it somewhat resembles *Diospyros obtusa* Ward,[2] from Sevenmile Creek, Montana, in the Fort Union group. The latter species is of approximately the same size and shape, but differs in the details of nervation. It is, however, quite close.

I have named this species in honor of Mr. Arnold Hague, of the United States Geological Survey.

[1] Cf. Lesquereux, Cret. and Tert. Fl., Pl. XXXIV, figs. 1, 2; Ward, Types of the Laramie Fl., Pl. XLIX, figs. 1, 2.

[2] Types of the Laramie Fl., p. 535, Pl. XLIX, fig. 5.

Habitat: Yellowstone River, one-half mile below the mouth of Elk Creek, base of bluff; collected by F. H. Knowlton, August, 1888.

OLEACEÆ.

Fraxinus wrightii n. sp.

Pl. XC, fig. 4.

Leaflet small, membranaceous in texture, oblong in outline, unequal-sided, wedge-shaped at base, obtuse at apex, margin with few irregular scarcely pointed teeth; midrib strong, slightly flexuose; secondaries about 7 pairs, alternate, at various angles, flexuose, camptodrome or subcraspedodrome, mostly arching and joining by bows some distance inside the margin, sometimes entering the teeth, and usually with outside branches to the minute, often obtuse, teeth; nervilles numerous, irregular, all forked or broken; finer nervation producing irregular quadrangular meshes.

The specimen figured, which was the only one found, is 4 cm. long and 2.2 cm wide. It is decidedly inequilateral, with a wedge-shaped base, and undulate-toothed margin. The nervation is camptodrome, with the secondaries arched and joined by broad bows well inside the margin, or occasionally with a secondary entering a tooth, thus becoming craspedodrome. The finer nervation is beautifully preserved, producing very irregularly quadrangular meshes.

The relation of this species is undoubtedly with *Fraxinus leerii* Lx.,[1] from Florissant, Colorado. Lesquereux's species differs in being much larger and narrower, with merely undulate margin. The nervation is strictly camptodrome, but otherwise identical.

I have named the species in honor of Mr. George M. Wright, by whom it was collected.

Habitat: Yellowstone River, below Elk Creek, top of bluff; collected by George M. Wright, September 9, 1885.

Phyllites crassifolia n. sp.

Pl. CII, fig. 5; Pl. CIII, fig. 4.

Leaves very large, thick, apparently rudely oval or orbicular in outline; base rounded or slightly heart-shaped, upper portion rounded(?); margin

---

[1] Cret. and Tert. Fl., p. 196, Pl. XXXIII, figs. 5, 6.

entire or undulate; petiole usually very thick (7 mm. in diameter); midrib thick (5 mm.), straight, splitting above into 2 equal branches; secondaries thick, straight, alternate or subopposite, often forking, craspedodrome or subcamptodrome, the secondaries or their branches united by broad loops with branches from the outside to the margin; nervilles very numerous, strong, mainly percurrent, yet often forked or broken; finer nervation, producing large, mainly irregular, quadrangular areolæ.

This species is based on a number of fragments that are insufficient to show the true character. Two of the largest are figured, showing what is assumed to be the base and upper portions. The largest is 13 cm. long and about 10 cm. wide, but this could have been only a fragment of the original size. This specimen (fig. 5 of Pl. CII) is peculiar in that the midrib splits in the upper portion into 2 equal branches, both of which are again branched on the outside. This leaf appears also to have been 2-lobed at the apex, all of which may be abnormal and due to an injury to the midrib. The nervation in the upper portion quite markedly camptodrome.

The lower portion that I have assumed to belong to this species has an exceedingly thick petiole, of which only a fragment is preserved, and also a thick midrib. They appear so different that it seems hardly probable that they can be identical, but rather than multiply unsatisfactory species they may remain as above until additional material can be obtained. None of the margin except the very base is preserved.

On account of the fragmentary nature of these leaves, I am unable to determine with satisfaction the proper genus to which they should be referred. In this uncertainty I have placed them provisionally under Phyllites.

Habitat: Cliff on west end of Fossil Forest Ridge; Fossil Forest Ridge, near head of Crystal Creek, various beds; collected by Ward and Knowlton, August, 1887, and by W. H. Weed, September 20, 1885.

CARPOLITHES OSSEUS Lx.

Carpolithes osseus Lx.; Ann. Rept. U. S. Geol. and Geog. Surv. Terr., 404, 1872 (1873).

A very doubtful species, of which the type is lost and the species not since obtained.

Habitat: "Elk Creek, near Yellowstone River; A. C. Peale, Joseph Savage, and O. C. Sloane."

CARPITES PEDUNCULATUS n. sp.

Pl. C111, fig. 3

Fruit round, apparently 4 or 5 celled; pedicel short, thick.

This fragmentary fruit is hardly worthy of description, for it may be so deformed by pressure that it can not be recognized again.

Among the described fruits of this heterogeneous class *C. dotani* Lx.) from Black Buttes, Wyoming, is perhaps closest, but probably the resemblance is only superficial.

Habitat: Yellowstone River, one-half mile below the mouth of Elk Creek, top of bluff (with Ulmus fruits); collected by F. H. Knowlton, August, 1888.

FOSSIL FORESTS

The fossil forests of the Yellowstone National Park are, beyond question, the most remarkable of their kind that have thus far been discovered in any part of the world. Isolated pieces or stumps of fossil wood are of common occurrence, being found in almost all quarters of the globe, from near the point farthest north that was reached by the Greely Arctic Expedition to southern South America; from Spitzbergen and Nova Zembla to South Africa and Australia, and geologically from the Devonian to beds in process of formation at the present day. In many localities there are aggregations of logs and stumps that are worthy to be dignified by the name of fossil forests; as, for example, in Chalcedony Park, near Holbrook, Arizona; near Calistoga, California, and in the vicinity of Cairo, Egypt. But in all of these places, so far as known, all or most of the trunks are prostrated and lie scattered about in the greatest confusion. In some cases there is evidence that the logs were transported by currents before being fossilized. The fossil forests of the Yellowstone National Park and vicinity, on the other hand, are not only more extensive in area, but the trees are almost all standing upright in the exact positions in which they grew originally. Many of these trunks, standing on the slopes and steeper hillsides, rise to a height of 20 or 30 feet, and are covered with lichens and blackened and discolored by frost and rain. At a short distance it is hard to distinguish them from the near-by living relatives. The following account by Prof

---

Foss. Fl., p. 365, Pl. LX, fig. 26.

W. H. Holmes, the discoverer of these fossil forests, shows the impression first made by the sight of them:

> As we ride up the trail that meanders the smooth river bottom, we have but to turn our attention to the cliffs on the right hand to discover a multitude of the bleached trunks of the ancient forests. In the steeper middle portion of the mountain face, rows of upright trunks stand out on the ledges like the columns of a ruined temple. On the more gentle slopes, farther down, but where it is still too steep to support vegetation, save a few pines, the petrified trunks fairly cover the surface, and were at first supposed by us to be the shattered remains of a recent forest.[1]

Fossil trees or fragments of wood of greater or less size are found in many parts of the Park, but their distribution is mainly confined to the northern and northeastern portions. The forests of standing trees are all found in the vicinity of the Lamar River, the most striking being exposed on the slopes and cliffs of Amethyst Mountain and Specimen Ridge. Nearly all of these forests are easily accessible from the well-traveled road between the Mammoth Hot Springs and the town of Cooke, Montana.

As the visitor enters the area drained by the Lamar River and by the smaller streams running into the Yellowstone below the Grand Canyon, evidences of proximity to the fossil forests are soon at hand. In the bed of every stream pieces of wood, often of considerable size, may be found. These pieces have in many cases been carried miles from their original source by the torrents incident to the melting of the snows in spring. In this way the pieces of wood have become rounded and worn and at remote distances are changed into smooth, rounded pebbles.

The first forest to be visited is near Yanceys, and is known as Yanceys Fossil Forest. It is located about 1 mile south of the hotel, on the middle slope of a hill that rises about 1,000 feet above the little valley. It is reached by an easy trail, and as one approaches, a number of trunks are observed standing upright among the stumps and trunks of living trees, and so much resembling them that a near view is necessary to convince the visitor that they are really fossil trunks. Only two rise to a considerable height above the surface. The larger one is about 15 feet high and 13 feet in circumference; the other is a little smaller. The roots are not exposed, so that it is impossible to determine the position of the part in view. Its original length can not, of course, be ascertained. It is also impossible to

---

[1] U. S. Geol. and Geog. Survey of the Terr. Hayden's Twelfth Annual Report, 1878 (1883), p. 48.

determine the original diameter, as the bark is in no case preserved. The standing trees are both conifers, and belong to the genus Cupressinoxylon.

Above these standing trunks many others are visible, but the disintegrating forces of nature keep them at about the same level as that of the surrounding rock, from the fact that they tend to break up easily into small fragments. Some of these trunks rise only a few inches, while others are nearly covered by the shifting débris. They vary in size from 1 to 4 feet in diameter, and are so perfectly preserved that the annual rings can be easily counted. The internal structure is also in most cases nearly as perfect as though the tree were living. The cells still retain their delicate markings, and often their perfect form.

There are numerous fossil leaves found in the rocks about the bases of these trees, but none apparently corresponding to the trunks; that is, the trunks are all coniferous, while the leaves are dicotyledonous; but from the nature of the case a coniferous trunk is much more readily preserved than a dicotyledonous one.

The next forest that claims attention is the one mentioned by Mr. Holmes, and is the one most frequently visited by observers. It is known locally as the Fossil Forest, and is exposed on the northern slope of Amethyst Mountain, opposite the mouth of Soda Butte Creek. The trunks may be easily seen from the road along the Lamar River and quite a mile away. They stand upright—as Holmes has said, like the pillars of some ancient temple—and a closer view shows that there is a succession of these forests, one above the other, through the entire 2,000 feet of this mountain. That is to say, in early Tertiary time a magnificent forest flourished in this region, which was buried under the débris ejected from volcanoes of greater or less size that are supposed to have existed in this vicinity. The trees were surrounded by silica-charged waters and were turned to stone. The area on which they grew was probably undergoing a very gradual submergence and the trees were slowly entombed. This is shown by the fact that the trees are in an upright position and were not broken by the incoming material which covered them.

After the first forest was entombed, quiet was restored for a sufficient length of time for a second forest to grow above it. Then volcanic activity was renewed, and the second forest was buried and silicified as the first had been. This process was repeated until 2,000 feet of volcanic material had been accumulated and not fewer than fifteen forests were entombed. Then the volcanoes ceased their activity and final quiet was restored. Probably an upward tendency was given to the area, but it must have been very gradual and not attended by the distortion which so frequently accompanies mountain building. The disintegrating action of frost and rain immediately set in and has carved out this mountain, in the heart of which may be read the story of its origin.

In the foothills and several hundred feet above the valley there is a per-pendicular wall of breccia, which in some places attains a height of nearly 100 feet. The fossil trunks may be seen in this wall in many places, all of them standing upright in the positions in which they grew. Their upright position proves that if there have been changes of level they have been gradual and in the same plane, as otherwise the trunks would be variously inclined. Some of these trunks, which are from 2 to 4 feet in diameter and 20 to 40 feet in height, are so far weathered out of the rock as to appear just ready to fall, while others are only slightly exposed. Niches mark the places from which others have already fallen, and the foot of the cliff is piled high with fragments of various sizes.

Above this cliff the fossil trunks appear in great numbers and in regular succession. As they are perfectly silicified they are more resistant than the surrounding matrix, and consequently stand out above it. In most cases they are only a few inches above the surface, but occasionally one rises as high as 5 or 6 feet.

The largest trunk observed in the Park is found in this locality. It is a little over 10 feet in diameter, which includes a portion of the bark. It is very much broken down, especially in the interior, a condition which very probably prevailed before fossilization. It projects about 6 feet above the surface.

The most remarkable of all the forests, however, is located on the west-ern end of Specimen Ridge, about 1 mile southeast of Junction Butte and opposite the mouth of Slough Creek. It was first brought to the notice, of the scientific world by Mr. E. C. Alderson and the writer, who discovered it in August, 1887. It is found on the higher portion of the ridge, and is several acres in extent. The trees are exposed at various heights on a very steep hillside, and the remarkable feature is that most of them project well above the surface.

One of the largest and best-preserved trees stands at the very summit of the slope. It is 26½ feet in circumference without the bark, and rises about 12 feet in height. The portion of this huge trunk preserved is the base, and below ground it becomes somewhat enlarged and passes into the roots, which are as large as the trunks of ordinary trees. The roots are embedded in the solid rock, as shown in the figure (see Pl. CIV).

This trunk is a true Sequoia, and is so closely allied to the modern redwood (*Sequoia sempervirens*) of California as to be hardly distinguishable from it. It would be interesting to learn the height this tree attained, but it seems safe to assume, from what we know of its living representative, that it must have been more than a hundred feet high.

Just below the large trunk, on the steep hillside, are two more standing trees (see Pl. CVI), which we may imagine to have formed the doorposts of the "ancient temple" of which Holmes speaks. They stand about 20 feet apart and rise about 25 feet in height. They are both about 2 feet in diameter and are also without the bark.

In other parts of the area there are standing trees which attain a height of 12 to 20 feet. They are all under 2 feet in diameter. In a few cases the bark is also preserved. It is hardly ever more than 3 inches in thickness.

Scattered about over the area are a great many trunks that rise only a few inches above the surface. These vary in diameter from 2 to 5 feet. They are often hollow in the center and have the cavity lined with brilliant amethyst crystals.

One of the larger trees appears to have been prostrated before it was fossilized (see Pl. CVIII). It is about 4 feet in diameter and is exposed for a length of 40 feet. There is nothing to indicate the portion of the trunk in its relation to roots and branches, but neither shows on the exposed part. There is no appreciable diminution in diameter, and consequently it must have been a very tall trunk.

The matrix about the bases of these trees, as well as those in the Fossil Forest, contains numerous impressions of leaves, branches, and fruits. In the Fossil Forest there are at least 6 horizons at which plant remains occur. These are separated by a few inches, or in some cases by many feet. In the forest last described, which may be called the Junction Butte Forest, there are only 2 or 3 plant horizons.

Most of the trunks in all three of the described forests are coniferous, but occasionally a dicotyledonous trunk is found, showing that the forest was to some extent a mixed one. It is of course more than probable that the leaves found in the matrix about the bases of the trees were at one time attached to them, but as they have never been found in association, it is manifestly impossible to correlate them.

The next fossil forest in rank of size is, perhaps, the one found on Cache Creek, about 7 miles above its mouth. It is exposed on the south bank of the creek, and covers several acres. The trunks are scattered from bottom to top of the slopes, through a height of probably 800 feet. Most of the trunks are upright, although there is only now and then one projecting more than 2 or 3 feet above the surface. The largest one observed was 6 feet in height and about 4 feet in diameter. While most of the trunks appear to the naked eye to be coniferous, there are a number that are obviously dicotyledonous. It is certain, however, that the conifers were the predominant element in this as in the other fossil forests.

The slopes of The Thunderer, the mountain so prominently in view from Soda Butte on the south, have also numerous fossil trunks. They are mainly upright, but only a very few are more than 2 feet above the surface. There were no remarkably large trunks observed, the average diameter being less than 2 feet.

Mount Norris, which is hardly to be separated from The Thunderer, has a fossil forest of small extent. The trees are of about the same size and characteristics as those on the larger mountain.

Forests of greater or less extent, composed mainly of upright trunks, are exposed on Baronett Peak, Bison Peak, Abiathar Peak, Crescent Hill, and Miller Creek. In fact, there is hardly a square mile of the area of this northeastern portion of the Park without its fossil forest—scattered trunks or erratic fragments.

The vast area to the east of the Yellowstone Lake has never been explored thoroughly from the paleobotanical side, but enough is known to be certain of the presence of more or less fossil wood. The stream beds contain occasional fragments, which is a sufficient indication that trunks of trees must be near at hand.

DESCRIPTION OF SPECIES.

SEQUOIA MAGNIFICA, n. sp.

Pls. CIV, CV, CX, CXI, CXVII, figs. 1-6.

Diagnosis.—Trunks often of great size, 6 to 10 feet in diameter, 30 feet high as now preserved, bark when present 5 or 6 inches in thickness, annual rings very distinct, 2 to 3 mm broad; fall wood reduced to narrow bands of 3 to 15 rows of thick-walled cells; cells of spring and summer wood large, hexagonal or often elongated; resin tubes numerous, composed of short cells; medullary rays numerous, of a single series or occasionally with a partial double series of superimposed cells; wood cells with one or two rows of small circular pits.

Transverse section.—In this section (Pls. CX, CXI) the structure appears beautifully preserved. The rings are rather narrow, being only 2 or 3 mm broad, or often only 1 mm. They are very sharply demarked, even to the naked eye. Under the microscope the rings are found to consist of a band of thick-walled cells that is never more than 15 rows of cells deep and often is reduced to 2 or 3 rows. The cells composing the spring and summer wood are of uniform size and inclined to be hexagonal in shape. Those of the fall wood are, of course, compressed.

The resin cells are numerous and may be readily distinguished by the dark contents. They occur mainly in the spring and summer wood.

The medullary rays seen in this section are long, straight, and separated by usually about 3 rows of wood cells.

Radial section.—This section (Pl. CXVI, figs. 2-3) is the least satisfactory of all. The wood cells show well under the microscope, but their markings are very obscure. By prolonged search it is made out that the pits are in 1 row, or sometimes 2 parallel rows. They are small, as far as can be made out, and are too obscure for satisfactory measurement.

The rays are composed of long, unmarked, cells.

Tangential section.—This section (See Pl. CXVI, fig. 1) is very satisfactory. The wood cells are long and unmarked. The resin ducts are numerous, but scattered, the cells being twice or three times as long as wide. In many cases they are filled with or contain masses of dark material, representing the resin now turned to a carbonaceous mass.

The medullary rays are composed of 1, or in some cases of a partially double, series of 2 to about 25 superimposed cells. They are large and quite thick walled. The average number of cells in each ray is about 12.

This species is closely related to the living *Sequoia sempervirens* Endl., more closely than any other fossil species with which I am familiar. They are hardly to be separated by any well-defined characters. The living wood has the same clearly marked annual rings, resin cells, partially double rays, and pits on the wood cells. The medullary rays in the living wood are provided with numerous round pores or markings. These seem to be absent from the fossil specimens, but, as already related under the diagnosis, the fossil is not well preserved in the radial section and they may have been present there when it was living. The dimensions of the various elements are much the same in the living and fossil specimens, thus leaving no doubt as to their close affinity.

In size of trunks these species are also similar. The largest trunks observed in the Yellowstone National Park belong to *S. magnifica*. They range in size from 4 to 10 feet in diameter, one of the largest being shown on Pl. CV. This is 26½ feet in circumference and stands upright on the hillside. It is 12 feet high, and represents the base of the trunk, as the large roots are well preserved. Their height is of course unknown, but one was fortunately prostrated before fossilization (Pl. CVIII), and is 40 feet long, with no apparent diminution in diameter. It is altogether likely that they may have been equal in height to some of the living representatives.

I have thought best to give this fossil species a name different from that of the living tree, notwithstanding the fact that they are evidently so closely related. The fossil comes from a locality remote geographically from the living redwood, and, moreover, from a horizon that, although comparatively recent, is so ancient as to make it extremely improbable that the type has actually been living for so long a period. There can, however, be no doubt that the living redwood is the direct descendant of this remarkable tree that was once so abundant in the Yellowstone National Park.

Habitat: Specimen Ridge, Fossil Forest at head of Crystal Creek, Fossil Forest on Cache Creek, etc.; collected by F. H. Knowlton, August, 1887–August, 1888.

PITYOXYLON ALDERSONI G. sp

Pls. CVI, CXII, CXIII, CXVIII, figs. 3, 4; Pl. CXIX, fig. 2.

Diagnosis.—Trunks of large size, 3 to 5 feet in diameter; annual rings very distinct, often 8 or 9 mm. broad, very sharply demarked; resin ducts numerous, large, scattered, occurring in late summer and fall wood; wood cells long, with a single irregular row of medium-sized pits; medullary rays in a single series, or occasionally with divided cells, rays from 2 to 25 cells high, the average being about 10 or 12 cells.

Transverse section.—The annual rings are very distinct, being plainly discernible to the naked eye. Some of the broadest rings are fully 9 mm. wide, and none are less than 6 mm. The demarcation between fall and spring wood is very pronounced (see fig. 4 of Pl. CXVIII and 2 of Pl. CXIX), the cells of fall being small, compressed, and thick-walled, while those of the early spring wood are very large, and, of course, thin-walled.

The cells of the spring and summer wood continue for a width of 5 mm., but little, if any, diminished in size. Then they become slightly smaller and thicker-walled and pass gradually into the fall wood.

The resin ducts are very large. They are not found in the summer wood, but occur irregularly in the early fall and late fall wood.

The medullary rays, as observed in this section, are straight and separated by from 3 to 8 or 10 rays of wood cells. The individual cells are apparently long.

Radial section.—Notwithstanding the fact that the wood seems to be perfectly preserved, it does not reveal the structure well in this section. The wood cells are seen to be sharp-pointed where they join. They are, of course, broad in the spring and summer wood, and very narrow and thick-walled in the fall wood. It is very difficult to make out the pits, but in exceptionally well preserved portions a few may be faintly seen. They are scattered, but in a single series. They are so obscure that no satisfactory measurements can be made.

The medullary rays in this section are long, thick-walled, and without markings, so far as can be made out.

Tangential section.—This section is very plain. The medullary rays are numerous and in a single series, although occasionally a ray may be observed in which there are 2 series of cells for a short distance. In such cases the cells are always smaller than the ordinary ray cells. The number of cells making up each ray ranges from 2 to 30 or more, but the average number is about 8 to 15.

The rays in which there is a resin duct are rather rare. The duct is large, taking up all the width of the ray. The remainder of the ray is 3 rows of cells high in the middle and is reduced to 1 at the extremities.

The wood cells show clearly in this section. They are not provided with pits or other markings.

Habitat: Specimen Ridge, Fossil Forest, near head of Crystal Creek; collected by F. H. Knowlton, August, 1887. Yancey Fossil Forest; collected by F. H. Knowlton, August, 1887.

PITYOXYLON AMETHYSTINUM n. sp.[1]

Pls. CVII, CVIII, CXIV, CXV, CXVIII, figs. 1, 2.

Diagnosis.—Trunks of small or medium size; annual rings sharply demarked, 3 to 8 mm. broad; resin ducts numerous, scattered, but mainly in fall wood; wood cells long, sharp-pointed, provided with a single row of scattered, small, somewhat irregular pits; medullary rays numerous, in a single series of 2 to 12 cells, the average being about 5 or 6.

Transverse section.—Much like the preceding species, except that the rings are narrower, the cells of spring and summer wood are smaller, and the late fall cells have thinner walls. The resin ducts are also much the same, being in general only a little smaller. A few are found in the summer wood, but most of them are in the fall wood. The rays are not nearly so numerous as in the last species. They are often separated by as many as 25 rows of wood cells.

Radial section.—The radial section of nearly all woods from the Yellowstone National Park is more or less obscure. The one under consideration is no exception to this rule, and it is only after considerable search that the pits can be determined. They are in a single row (see Pl. CXVIII, fig. 1)

---

[1] In 1888 Dr. J. Felix, of Berlin, visited and collected fossil wood in the Yellowstone National Park. The results of his work were published in Zeitschrift der Deutschen geologischen Gesellschaft, for 1896. He described six species of fossil wood, of which number I have recognized four. The following two species were not figured, and as the locality whence they came is more or less in doubt I have not included them in the systematic enumeration. They are as follows: *Pityoxylon fallax* and *Cupressinoxylon cuterion*. They may be identical with certain of the species I have described, but of this I am uncertain.

and are rather small. They are so obscure that it is impossible to make trustworthy measurements.

The medullary rays, as seen in this section, are composed of long, thin-walled cells, and so far as can be determined they are without pits or other markings.

Tangential section.—This section (Pl. CXVIII, fig. 2) shows the structure very plainly. The medullary rays are abundant and always in a single series, except the large compound ones. The number of cells in each ray varies from 2 to 10 or 12, the average number being about 6. The compound rays inclosing the resin ducts are rather small, with three rows of cells in the middle portion. No markings can be made out on the wood cells in this section.

This species is very closely allied to the one preceding, and should perhaps be referred to it. The main points of difference are the following: Narrower annual rings; smaller resin ducts, that are occasionally found in the summer wood; smaller wood cells throughout; smaller and shorter compound medullary rays; ordinary rays always in a single series of 2 to 12 cells (average 6) instead of from 2 to 30 or more (average 12).

Habitat: Specimen Ridge, Fossil Forrest, near head of Crystal Creek; collected by F. H. Knowlton, August, 1887.

LAURINOXYLON PULCHRUM n. sp.

Pls. CXVI, CXIX, figs. 3-5; Pl. CXX, fig. 1.

Transverse section.—Annual ring very distinct to the naked eye, 2 to 4 mm. broad. The demarcation between the rings results from 10 or 12 layers of thicker-walled cells, representing the late fall wood, and from the greater abundance of ducts in the immediately following spring wood.

The wood cells are small and arranged in serial rows except in the vicinity of the ducts, where they are somewhat irregular (see fig. 1 of Pl. CXX). Surrounding the ducts, and sometimes filling the remainder of the space between rays, the cells are larger and not so completely seriated. The ordinary wood cells are about 0.01 mm. in diameter, and those near the ducts 0.015 or 0.02 mm. There is an occasional row of the large-sized wood cells along a ray, as in fig. 1 of Pl. CXX.

The ducts are very plainly shown in this section. At least half of them are single and nearly or quite circular in section. Of the remainder, most are double, while occasionally there are 3 in a row or series, and exceptionally as many as 4. The number in a square millimeter is only from 4 to 6. The smallest of the single ducts range in diameter from 0.03 to 0.06 mm. The largest observed are 0.24 mm. in long diameter and 0.16 mm. in short diameter. The largest double duct—that is, when there are two together—is 0.34 mm. The largest of the series of 3 is 0.36 mm., and the largest of the few in series of 4 is 0.44 mm. The average diameter of large and small ducts is probably about 0.12 mm.

The medullary rays are 1 to 3 cells wide, and run irregularly among the ducts. They are about 0.01 mm. broad.

Radial section.—The wood cells are long, slender, and apparently sharp-pointed. There is evidence also that some of these are divided up into short cells by square divisions.

The medullary rays form plates of short, rather thin-walled cells. They are from 0.02 to 0.04 mm. in diameter and from 0.05 to 0.09 mm. in length. They do not appear to be marked, yet there is some evidence that there were minute pits; but the specimens are not well enough preserved to be certain of this.

The ducts shown in longitudinal section (Pl. CXIX, fig. 3) are very pronounced. The individual cells are from 0.10 to 0.20 mm., or sometimes more, in length. The walls are covered with small round pits, which occasionally pass into regular scalariform markings (see figs. 4 and 5 of Pl. CXIX.) Each duct is surrounded by a mass of tissue from 2 to 6 or 8 layers of cells thick, of which mention was made under the discussion of the transverse section. The individual cells of this sheath are of about the same size and appearance as the large cells of the medullary rays.

Tangental section.—The fine photomicrographic reproductions of this section (Pl. CXVI) give a far better idea of the structure than any description can. The medullary rays, it will be observed, are very numerous (about 3 to each square millimeter). They are from 1 to, exceptionally, 4 layers of cells broad and about 12 layers high, the extremes being 5 and 20.

This plate shows admirably the ducts and related tissue. The one in the center of the plate shows well the manner of division, although the magnification is hardly sufficient to show the pits or markings.

This species is one of the handsomest with which I am familiar. It has affinities with a number of described forms, as, for example, *Laurus triseriata* Caspary,[1] from the Tertiary of Prussia. From this it differs in the arrangement of ducts and in rays, and somewhat in the markings. It is also evidently allied to the two forms from the Tertiary of Arkansas, *Laurinoxylon branneri* Kn.[2] and *L. bispiratum* Kn.

The genus Laurus was evidently abundant in this flora, and it is to be expected that the trunks would be occasionally preserved. It is of course probable that this wood may belong to a species that has also been described from the leaves, but there is manifestly no means of connecting them.

Habitat: Specimen Ridge Forest, near head of Crystal Creek, Yellowstone National Park, a prostrate log; collected by F. H. Knowlton, August 25, 1887.

PERSEOXYLON AROMATICUM Felix.

Perseoxylon aromaticum Felix: Untersuchung über fossile Hölzer, v. Stück; Zeitschr. d. Deutsch. geol. Gesell. Jahr. 1896, p. 254, 1896.

Laurinoxylon aromaticum Felix: Die Holzopale Ungarns, p. 27, Pl. I, fig. 7; II, fig. 7, 9.

This species was detected by Felix in his visit to the Yellowstone National Park in 1888. I did not meet with it.

Habitat: Vicinity of Yanceys, Yellowstone National Park. Collected by J. Felix, 1888.

PLATANINIUM HAYDENI Felix.

Pl. CXX, figs. 3–5.

Plataninium haydeni Felix: Untersuchung über fossile Hölzer, v. Stück; Zeitschr. d. Deutsch. geol. Gesell. Jahr. 1896, p. 254, 1896.

Transverse section.—The annual rings are faint, yet they may be seen with the naked eye. They are about 2 mm. broad. The medullary rays are very distinct in the weathered specimen.

Under the microscope the structure is shown to be well preserved. The wood cells are not arranged in radial rows, but are quite irregularly placed. They are large (0.01 to 0.03 mm.) and angular, being 5 to 6 sided by compression.

---

[1] Einige foss. Hölzer Preussens: Abhandl. z. geol. Specialk. v. Preussen u. Thüringischen Staaten. 1889, p. 60, Pl. XI, figs. 6–12; Pl. XII, figs. 1–5.

[2] Fossil woods and lignites of Arkansas: Ann. Rept. Geol. Survey Arkansas, 1889, Vol. II, p. 256, Pl. IX, figs. 8–9; Pl. X, figs. 1, 2; Pl. X, fig. 4.

[3] Op. cit., p. 258, Pl. X, figs. 3, 4; Pl. XI, figs. 3, 4.

The ducts (Pl. CXX, fig. 4) are very numerous. They occupy at least one-third of the area, exclusive of the rays. They are almost always single, although often placed close together, especially in the beginning of the spring wood. They are uniformly oblong in shape. In the spring wood the average size is 0.09 mm. in long and 0.06 mm. in short diameter. In the fall wood they are from 0.03 to 0.06 mm. in long and 0.025 to 0.05 mm. in short diameter.

The annual ring consists of a layer of slightly thicker wood cells, but it is mainly distinguished by the abruptly larger ducts in the spring wood (see fig. 4 of Pl. CXX).

The medullary rays are very abundant as seen in this section. They are from 1 to 10 or 15 cells broad. Fully 30 per cent of the area is covered by the medullary rays. The rays uniformly contain a black carbonaceous substance, these making them stand out in bold relief.

Radial section.—The most prominent feature in this section (Pl. CXX, fig. 5) is the medullary rays. They form high plates of usually short cells with black carbonaceous contents. The ducts are also prominent, and appear to be marked with scalariform thickenings, but as they are quite obscure, this is not positive.

Tangential section.—The structure of this section is very clearly revealed under the microscope. The medullary rays are very numerous. They range from 1 to 10 or 15 layers of cells broad and more than 100 high. The cells are round, thin-walled, and usually or not at all compressed. They take up, as already stated, fully 30 per cent of the space. In some cases the rays are 0.5 mm. long and 0.35 mm. broad (cf. fig. 3 of Pl. CXX).

The wood cells are long and sharp-pointed. So far as can be made out, there are few if any square divisions of the cells.

The ducts, of course, show well in this section, but the markings, if present, are now obscure.

This species is quite closely related to the living *Platanus occidentalis* L., the common sycamore or plane tree. The living wood shows the indistinct annual ring, the irregular wood cells, and numerous medullary rays almost identical with the fossil wood. There are certain minor points of difference, such as markings on the rays, lignification of the ducts, etc., but they are certainly close enough to make their generic identity reasonably sure.

The fact that Platanus leaves are very abundant in the beds surrounding the fossil trunks makes it extremely probable that the generic reference is correct. It is of course also probable that some of the leaves belong to the wood here described as different, but as they have never been found attached, it is manifestly unsafe to assume that there was ever organic union.

A number of fossil species have been described from various parts of the world; none, however, from North America. The general agreement between these and the one under consideration is close, but the specific differences are marked in certain cases. One of the nearest forms is *Platanus klebsii* Casp.,[1] from the Tertiary of Prussia. It differs in important minor characters, as does *P. borealis* Casp.,[2] from the same place. The two species described by Felix, *Plataninium parvum* Felix and *P. regulare* Felix, have only general resemblance.

In the original MS., which was submitted in March, 1896, I had of course given this another specific name, and it may still prove to be different from the *P. lanterai* of Felix. Unfortunately Felix has not figured his species, and it is difficult, from a mere technical description, to be entirely certain of their identity. It is reasonably certain, however, that they are identical, and I have so regarded them.

Habitat: Specimen Ridge Forest, near head of Crystal Creek, Yellowstone National Park. From a trunk 6 inches in diameter and about 1 foot in height; collected by F. H. Knowlton, August 25, 1887.

RHAMNACINIUM RADIATUM Felix.

Pl. CXVII, figs. 6, 7; Pl. CXIX, fig. 1.

Rhamnacinium radiatum Felix: Untersuchung über fossile Hölzer: Zeitschr. d. Deutsch. geol. Gesell., Jahr. 1896, p. 252, Pl. VI, fig. 3, 1896.

Transverse section.—Annual ring broad (7 mm.), very indistinct, consisting of only 1 or 2 rows of slightly thickened wood cells and rather abrupt presence of numerous large ducts in succeeding spring wood. Ducts very numerous, in radial rows. A few of the ducts are single, but mainly they are contiguous, with 2 to 10 in a series. The usual number is 5 or

---

Einige foss. Hölzer Preussens: Abhandl. z. geol. Specialk. v. Preussen u. Thuringischen Staaten, 1889, Pl. VIII, figs. 1-21.

Op. cit., Pl. IX, figs. 1-11.

four. The ducts occupy nearly one-half of the area, thus producing an open, soft wood. The longest series of ducts, embracing 10, is 0.50 mm. in length. Series of 4 or 5 having a length of 0.50 mm. are common. The small single ducts are 0.05 to 0.07 mm. in long and 0.04 to 0.05 mm. in short diameter. The average short diameter of all ducts is about 0.07 or 0.08 mm.

The wood cells are arranged in distinct radial rows. They are rather large and thin-walled, also showing that the wood was a soft, porous one.

The medullary rays in this section are rather numerous. They are 2 or sometimes 3 cells wide, and the cells are short and thin-walled.

Radial section.—The ducts appear especially numerous in this section. The marking on the walls is rather obscure, but they seem to be uniformly provided with minute pits.

The rays form high plates of short, thin-walled cells, apparently with small circular or oblong pitlike markings.

The wood cells are very long. They have sharp-pointed extremities and thin walls.

Tangential section.—This section is very characteristic, the most prominent feature being, of course, the cut off ends of the medullary rays. The rays are various, being 2 or rarely 3 or 4 layers of cells wide. The number of vertical rows is very indefinite, being rarely less than 10 or more than 30. The cells are rectangular, being often twice as long as wide. Some of the cells in the middle of the ray are more or less irregular in shape. All are very thin-walled.

The wood cells are the same as in the radial section.

The ducts are also prominent. They have oblique partitions and the walls are provided with round pits. The markings on the walls are not different from those to be observed in the radial section, but they happen to be better preserved.

In my original MS. this form was described under the new generic name of Populoxylon, from its undoubted close resemblance to wood of living Populus. It is with some hesitation that I transfer it to Felix's species, for they do not agree in every particular. On the whole, however, it is more than probable that they are the same, and I have so regarded them. The generic diagnosis, based upon the wood from the Park only, may be drawn up as follows: Annual ring present, but faintly demarked; wood cells long, narrow, sharp-pointed, thin-walled; ducts very numerous, occupying about one-half of the area, in radial rows of from 2 to 10, pitted, the pits small, round; medullary rays numerous, of short, thin-walled cells, rectangular or irregular in transverse section, arranged in 2 to sometimes 4 vertical rows of approximately 10 to 30 cells each.

Habitat: Specimen Ridge, near head of Crystal Creek, Yellowstone National Park; collected by F. H. Knowlton, August 22, 1887.

QUERCINIUM LAMARENSE n. sp.

Pl. CXVIII, fig. 5; Pl. CXX, fig. 2; Pl. CXXI, figs. 1, 2.

Transverse section.—Annual ring present, but very faint; consisting of but 1 or 2 rows of thickened cells. In the succeeding spring wood the ducts are much larger, thus making the ring visible to the naked eye.

Ducts numerous, scattered, most abundant in spring and summer wood; all single—that is, not contiguous. They are almost perfectly circular, being very slightly elongated radially. They are large, though not remarkably so for the genus; the larger ones ranging in diameter from 0.16 to 0.23 mm., the smaller being about 0.20 mm. The very smallest ducts are 0.05 mm. in diameter, and the more common of the small ones are 0.10 or 0.12 mm. in diameter. None of the ducts are arranged in notable radial rows.

The wood cells are in distinct radial rows, and are large and thick-walled. In most the lumen is nearly obliterated. The average size of the wood cells is 0.02 mm.

The medullary rays are neither very numerous nor conspicuous. They are mainly only 1 cell broad, with an occasional wide one of 20 or more cells, as will be described under the tangential section. Some of the single-celled rays pass for a considerable distance among the ducts, but by far the larger number lie between two ducts (see fig. 1 of Pl. CXXI).

Radial section.—The only sections available in this direction were, unfortunately, from poorly preserved portions of the specimen, and do not show the structure clearly. The wood cells, so far as can be made out, are very long, and, as shown by the transverse section, have thick walls. The rays form high plates of cells, the exact length of which can not be determined with satisfaction. If there were markings on the rays they can not be seen; neither can the markings on the ducts be observed.

Tangential section.—This section shows much better under the microscope than the radial one.

The rays are found to be of two distinct kinds: The most numerous are only 1 cell broad and from 10 to 25 cells high, the individual cells being thin-walled and oblong in shape. At scattered intervals are very broad rays composed of 10 to 20 rows of cells and extending for long distances through the section (see fig. 2 of Pl. CXXI). These broad rays are often somewhat cut by wood cells passing diagonally through them (see fig. 2 of Pl. CXXI). This does not, however, interfere with the ray as a whole, which is clearly demarked from the small rays of a single series of superimposed cells. The individual cells of the large rays are nearly circular in cross section, or more or less 6-sided by mutual pressure. They are also thin-walled.

Associated with the small rays is usually a layer or two of short-celled tissue or series of parenchymatous cells. Except for there being shorter cells they are not to be distinguished from the ordinary wood cells.

The ducts show clearly enough in this section, but they are not well enough preserved to permit the markings on the walls to be made out. It would seem that the walls were pitted, but this is largely surmised.

A considerable number of species of *Quercinium*,[1] or oak wood, in a fossil state, have been described from various parts of the world. Wood of this kind is readily distinguished by the large isolated ducts and the two kinds of medullary rays.

The species under consideration resembles a number of described forms, but they are all from the Old World, and are readily distinguished from it.

This species is closely allied to *Quercinium knowltoni* Felix, and may possibly be the same, but as Felix's species is not fully illustrated it is difficult to be positive. *Q. knowltoni* seems to differ in the shape and size of the large ducts, but it will need a careful comparison of the sections to be positive. For the present, at least, they may remain distinct.

Habitat. Specimen Ridge, Yellowstone National Park; specimen from

---

Fifteen species and varieties.

an upright trunk, 1 foot in diameter; collected by F. H. Knowlton, August 22, 1887.

QUERCINIUM KNOWLTONI Felix.

Quercinium knowltoni Felix: Untersuchung über fossile Hölzer: Zeitschr. d. Deutsch. geol. Gesell., Jahr. 1896, p. 250, Pl. VI, fig. 2, 1896.

As stated under the preceding species, these 2 forms may be identical, but in absence of full drawings of *Q. knowltoni* it seems best to regard them as distinct. The size and shape of the ducts certainly differ greatly.

Habitat: Amethyst Mountain, Yellowstone National Park, collected by J. Felix in 1888.

BIOLOGICAL CONSIDERATION OF THE TERTIARY FLORA.

The Tertiary flora of the Yellowstone National Park possesses great biological interest. It is a rich flora, and on comparing it with the living flora it becomes apparent that great climatic changes must have taken place since the close of the Miocene period to have made these modifications in plant life possible. The fossil flora embraces about 150 forms that have been distributed among 53 natural families. Following is a list of these families, with the number of species or forms referred to each:

	Species		Species
Fibers	10	Platanaceæ	5
Equisetaceæ	4	*Leguminosæ*	5
Coniferæ	13	Anacardiaceæ	4
Typhaceæ	1	Celastraceæ	4
Sparganiaceæ	1	*Ilicineæ*	2
Cyperaceæ	4	Sapindaceæ	5
Smilaceæ	1	*Rhamnaceæ*	4
Musaceæ	4	Vitaceæ	4
Juglandaceæ	8	Sterculiaceæ	4
Myricaceæ	3	Credneriaceæ	4
Salicineæ	10	Tiliaceæ	2
Betulaceæ	2	Araliaceæ	6
Fagaceæ	15	*Cornaceæ*	2
Ulmaceæ	5	*Ericaceæ*	4
Urticaceæ	10	Ebenaceæ	5
Magnoliaceæ	5	Oleaceæ	4
Lauraceæ	12	Phyllites, Carpites	5

The orders that are also found in the present flora are printed in italics.

The excellent 'Flora of the Yellowstone National Park,' by Mr. Frank Tweedy, has been made the basis of all comparisons between the fossil and living floras. According to Tweedy, the present flora embraces 69 natural families, 273 genera, and 657 species. The fossil flora embraces 33 families, 63 genera, and 148 species. The living flora has, therefore, 4 genera to each order and 2.4 species to each genus, while the fossil flora has not quite 2 genera to each family and 2.3 species to each genus. The relative proportion between the families, genera, and species is shown to be approximately the same in the Tertiary and the living floras. A still further comparison shows that there are a fraction more than twice as many living as fossil families, 4.3 times as many living genera, and 4.6 times as many species.

On comparing the families in the two floras, it is found that 19 of the 33 fossil families are not represented in the living flora. In the list of families above given the ones not italicized are the families not represented at the present time. It will be seen that such important families as the Juglandaceæ, Fagaceæ, Ulmaceæ, Magnoliaceæ, Lauraceæ, Platanaceæ, Anacardiaceæ, Celastraceæ, Vitaceæ, Sterculiaceæ, Tiliaceæ, Araliaceæ, Ebenaceæ, and Oleaceæ are not represented in the present flora. In other words, there are no walnuts, beeches, oaks, chestnuts, elms, magnolias, sycamores, sumacs, grapes, lindens, aralias, persimmons, or ashes at the present day. The absence of such important trees and shrubs produces a profound modification of the floral surroundings.

The dominant element in the living flora consists of the abundant coniferous forests; yet only 8 species are represented, and of these only 5 are at all common, and 65 per cent of the whole coniferous growth is made up of 4 species. The fossil flora is represented by 13 species, or nearly twice as many as the living. Among them was a magnificent Sequoia that was closely allied to the living *Sequoia sempervirens* of the Pacific coast. It had trunks 10 feet in diameter and probably of vast height. There were also 2 well-marked species of Sequoia, known from the leaves, and a number of supposed Sequoia cones. The pines were also abundant, no fewer than 8 species having been detected.

The deciduous leaved trees and shrubs of the Yellowstone National Park are conspicuously few in numbers. There are 2 species of Betula, 2 of Alnus, 7 of Salix, 2 of Populus, 1 of Acer, 1 of Vaccinium, 5 of the order Caprifoliaceae, 2 of Cornaceae, 2 of the Rosaceae, etc. Perhaps the most conspicuous tree is the quaking aspen (*Populus tremuloides*). The cottonwood (*P. angustifolia*) is rare, being found only along Cache Creek. Several of the willows are abundant, as is also the common birch (*Betula glandulosa*), and the June berry (*Amelanchier alnifolia*). The other shrubs are rare, or are confined to few localities.

The fossil flora, on the other hand, was especially rich in deciduous leaved vegetation. Thus the Juglandaceae was represented by 5 species of Juglans and 4 species of Hicoria (Carya), a number of which were very abundant. The genus Populus was especially rich, there being no fewer than 7 species. Certain of these, as *Populus speciosa*, *P. daphnogenoides*, and *P. glandulifera*, were in great abundance, and the stratum in which they occur consists of a perfect mat of these leaves. Something like 100 examples of 4 species were obtained.

Another striking feature was the presence of numerous magnificent magnolias. Of these, 4 species have been described from the leaves and 1 from the thick petals of the flower. The species described as *Magnolia spectabilis* is represented by a great number of leaves in a fine state of preservation. It appears to be more closely related to the living *M. grandiflora* (*M. foetida* of later authors) than any one previously described.

The sycamores were also an important element in this flora. Of the 2 species described from the leaves and 1 from the wood, the one known as *Platanus guillelmae* was especially abundant. It is found in nearly all the Tertiary beds in the Park and is represented in the collections by nearly 200 examples. The species described as *Plataninium haydeni* is based upon a trunk or branch 6 inches in diameter. It is most closely related to the living *Platanus occidentalis*.

Another important group is formed by 4 species of Aralia. Of these, *Aralia notata* was evidently one of the most abundant and imposing trees of the whole flora. The collections contain over 100 examples, none of which are entire, however, as some of the leaves must have been fully 3 feet in length and more than 2 feet in width. A small leaf and one of medium size are figured on the plates. *Aralia whitneyi*, a species common to the Auriferous gravels of California, had striking 5 to 7 lobed leaves, often 1 foot in length. This species was not so abundant, judging from the fossil remains, as the former species, but it was apparently quite widely distributed. The other species had smaller 3 or 5 lobed leaves.

The family Lauraceae was strongly represented by 5 genera, 11 species, and a large number of examples. The genus Laurus, which is now exclusively an Old World group, was represented by 6 well-marked species. The genera Malapoenna or Litsea and Cinnamomum, other Old World forms, were both represented, the former by 2 and the latter by 1 species. The genus Persea, an extensive Old World genus, with species also in tropical America and the southern United States, was represented by 1 species, which is closely related to a small tree now living in the South.

Another large and important group, now entirely unrepresented in the Park, is the Fagaceae, embracing 2 species of Fagus, 1 of Castanea, 11 of Quercus, and 1 of Dryophyllum. The Fagus here described is a beautiful, characteristic leaf and was evidently rare, as only a few examples were obtained. The Castanea, on the other hand, was very abundant and widely distributed within the Park. The leaves are large, and as handsome and striking as are the leaves of the living species. The oaks, however, were abundant in species and usually in individuals, and all but 3 proved to be new to science. Perhaps the most marked are *Quercus guiuea*, *Q. calvera*, and *Q. mossidentata*.

The family Urticaceae, which is represented in the living flora by a single rare herb (*Urtica gracilis*), was represented during Tertiary times by some 10 species of Ficus and a single more or less doubtful species of Artocarpus. Several of the figures are represented by a large number of specimens—as, for example, *Ficus densifolia*—but most of them were rare, at least as evidenced by the fossil remains. It is of great interest to learn, however, that they were once present in a region that has long since ceased to support them. The curious leaf referred provisionally to Artocarpus is also of much interest as indicating the possible presence of the bread-fruit trees in this portion of the American Continent. Two unmistakable species of Artocarpus have already been detected, 1 from the Laramie and Denver beds of Colorado, and the other from the Auriferous gravels of California and the Miocene of Oregon. It is therefore not improbable that this type was in existence in the Yellowstone National Park during the early Tertiary.

The family Leguminosae, now represented by a host of small herbaceous plants, was then represented by 3 species of Acacia and 2 of Leguminosites, but the fossil forms are not particularly satisfactory. The forms referred to Acacia consist of well-defined pods and are somewhat conventionally regarded as representing the modern Acacia. No leaves were obtained that could with satisfaction be held as representing the foliage of these pod-bearing shrubs or trees. The 2 species of Leguminosites are supposed to represent leaflets of some leguminous plant, but beyond this it is not possible to venture.

The only remaining group of deciduous-leaved plants of any magnitude is the Sapindaceæ, with 5 species of Sapindus. One of these, *Sapindus affinis*, is perhaps the most abundant form found among the Tertiary plants. The small characteristic leaflets are found in the greatest profusion. The other species were less abundant.

The other forms that require mention are: Ulmus, 4 species; Acer, at least 2 species, Celastrus, 3 species, and Rhamnus, Paliurus, Zizyphus, Cissus, Pterospermites, Tilia, and Rhus, with a single species each.

The vascular cryptogams appear to have been a more prominent feature of the flora during Tertiary times than at present. Of the 2 families present, the Filices and Equisetaceæ, the former is represented by 10 and the latter by 4 species, while the living flora has but 6 ferns and 4 horsetails, all rare.

The ferns were evidently abundant. They belong to 6 genera, and are represented in several cases by a large number of specimens. The largest genus is Asplenium, with 4 species. The species described as *Asplenium magnum* is one of the largest and finest forms that has been detected outside of the Carboniferous. *Asplenium idahoense* is also a large, well-marked species. The genus Dryopteris, the old Aspidium, is represented by 2 species, both of which are rather rare. They are, however, both fruiting, a condition of uncommon occurrence among fossil forms. There is also a beautiful Woodwardia, quite closely allied to a species now living in the eastern United States, and fine examples of the widely distributed climbing fern (*Lygodium kaulfussi*). The only living North American species (*L. palmatum*) is found from Massachusetts and New York south to Kentucky and Florida, and is generally rare throughout its range. The other ferns are an Osmunda and a delicate form referred provisionally to the genus Devallia.

The genus Equisetum, although represented by 4 more or less satisfactory species, was not abundant or particularly important. The most abundant form (*E. bomeri*) is small and has much the appearance of the living *E. hiemale*. The largest form (*E. microdontum*) is very rare. It was about 5 cm. in diameter.

From what has been presented, it is obvious that the present flora of the Yellowstone National Park has comparatively little relation to the Tertiary flora, and can not be considered as the descendant of it. It is also clear that the climatic conditions must have greatly changed. The Tertiary flora appears to have originated to the south, while the present flora is evidently of more northern origin. The climate during Tertiary time, as made out by the vegetation, was a temperate or subtemperate one, not unlike that of Virginia at the present time, and the presence of the numerous species of Ficus would indicate that it even bordered on subtropical. The conditions, however, that permitted the growth of this seemingly subtropical vegetation may have been different from the conditions now necessary for the growth of these plants. Thus, the genus Dicksonia is at present a tropical or subtropical genus, yet at least 1 species is distributed well into the temperate region. If a series of beds should be discovered in which there were a large number of Dicksonias, it might be supposed to indicate tropical or subtropical conditions; yet, as a matter of fact, these species may at that time all have been so constituted as to grow in a temperate land, and the genus as a whole may have become tropical in recent times. Following out this general line of argument, it may be said that while the Tertiary vegetation of the Yellowstone National Park would, from our present standard, be regarded as indicating a temperate or possibly warmer climate, the actual conditions then prevailing may have been quite different. It is certain, however, that the conditions were very different from those now prevailing.

Table showing the distribution of the Tertiary plants of the Yellowstone National Park.

	Distribution in the Park														Distribution outside			
	Fort Union (Eocene)			Intermediate (Miocene)					Lower Flora (Miocene)									
Species	Yellowstone River, half a mile below mouth of Elk Creek.	Yellowstone River, 1 mile below Elk Creek, opposite Wednesday Creek.	Southeast side Cross at Hill	Ground Hill above Creek, Yancey.	Peak west of Dunraven, on Cottonwood Creek.	Lamar Creek.	Fossil Forest and Yancey's, 1 mile south of Yancey.	Hill south of Yancey.	Hill above fourth of Lamar Creek.	Specimen Ridge near crystal Creek.	Fossil Forest.	Cliff west of Fossil Forest near Chalcedony Creek.	Lamar River, between Cache and Calfee creeks.	Lamar (red beds).	Bozeman and Livingston.	Fort Union group.	Green River group.	Auriferous gravels, California.
	1.	2.	3.	4.	5.	6.	7.	8.	9.	10.	11.	12.	13.	14.	15.	16.	17.	18.
1. Woodwardia procerolata n. sp											6							
2. Asplenium iddingsi n. sp											6							
3. Asplenium magnum n. sp																		
4. Asplenium crossun (Lx.)														b				
5. Asplenium remotidens n. sp																		
6. Dryopteris weedii n. sp																		
7. Dryopteris xantholithensis n. sp											6							
8. Davallia ? montana n. sp											3							
9. Lygodium kaulfussi Heer																		
10. Osmunda affinis Lx.								1										
11. Equisetum hagaci n. sp								5										
12. Equisetum lesquereuxi n. sp																		
13. Equisetum canaliculatum n. sp																		
14. Equisetum deciduum n. sp											7							
15. Pinus gracillistrobus n. sp											7							
16. Pinus premurrayana											7							
17. Pinus macrodepis n. sp																		
18. Pinus sp																		
19. Pinus wardii n. sp											4							
20. Pinus iddingsi n. sp																		
21. Taxites olriki Heer																		
22. Sequoia couttsiae Heer																		
23. Sequoia langsdorfii (Brgt.) Heer					6		3		6, 7									
24. Sequoia, remains of											5, 6							
25. Phragmites? latissima n. sp																		
26. Sparganium stygium Heer																		
27. Cyperacites angustia Al. Br. ?																		
28. Cyperacites giganteus n. sp																		
29. Cyperacites sp																		
30. Cyperacites sp																		
31. Smilax lamarensis n. sp											6						×	
32. Musophyllum complicatum Lx.																		
33. Juglans californica Lx.											6						×	×
34. Juglans rugosa Lx.						4	5, 6, 9, 7								×		×

a The numbers refer to the beds in which the plants were found.
b Cherry Creek, Oregon.
c Rare.
d Miocene: Elk Creek, Lx.

Table showing the distribution of the Tertiary plants of the Yellowstone National Park—Continued.

	Distribution in the Park.													Distribution outside.			
	Fort Union (Eocene).						Intermediate (Miocene).			Lamar Flora (Miocene).							
Species.	Yellowstone River, half a mile below mouth of Elk Creek.	Yellowstone River, 1 mile below Elk Creek, opposite Hellroaring Creek.	Northeast side Crescent Hill.	Crescent Hill above twenty Yancey.	Peak west of Bugaston, east of Carnelian Creek.	Tower Creek.	Fossil Forest and vicinity, 1 mile south of Yancey.	Hill south of Yancey.	Hill above fourth of Lost Creek.	Specimen Ridge, near Crystal Creek.	Fossil Forest. a	Cliff west of Fossil Forest, near Chalcedony Creek.	Lamar River, between Cache and Calfee creeks.	Lamamie Creek (Laramie).	Beartee and Lavalization.	Fort Union group.	Green River group.	Auriferous gravels, California.
	1.	2.		4.	5.	6.	7.	8.	9.	10.	11.	12.	13.	14.	15.	16.	17.	18.
35. Juglans schimperi Lx.																	x	
36. Juglans laurifolia n. sp.																		
37. Juglans crescentia n. sp.											6							
38. Hicoria antiqua (Newby.)											6							
39. Hicoria crescentia n. sp.																		
40. Hicoria culveri n. sp.																		
41. Myrica scottii Lx.																		
42. Myrica wardii n. sp.											5							
43. Myrica lonarensis n. sp.																		
44. Populus glandulifera Heer.		?																
45. Populus speciosa Ward																		
46. Populus xantholithensis n. sp.																		
47. Populus daphnogenoides Ward																		
48. Populus balsamoides Goepp.																	x	
49. Populus ? vivaria n. sp.																		
50. Salix varians Heer.																		x
51. Salix angusta Al. Br.																		
52. Salix lavateri Heer.																		
53. Salix elongata ? O. Web.									10	L. 8.								
54. Betula bldingsi n. sp.																		
55. Corylus macquarryi Heer.										M. 8.						?	x	
56. Fagus antipofi Heer.												S?					?	
57. Fagus undulata n. sp.																		
58. Castanea pulchella n. sp.											?							
59. Quercus grossidentata n. sp.											5							
60. Quercus consimilis ? Newby.																		x
61. Quercus? magnifolia n. sp.																		
62. Quercus furcinervis americana											5							x
63. Quercus weedii n. sp.										M. 8. 6								
64. Quercus sp.																		
65. Quercus chlorsii Heer.													1			?		x
66. Quercus yancyi n. sp.																		
67. Quercus culveri n. sp.																		
68. Quercus hesperia n. sp.																		

a The numbers refer to the beds in which the plants were found.   b Miocene of Alaska

Table showing the distribution of the Tertiary plants of the Yellowstone National Park—Continued.

		Distribution in the Park.										Distribution outside					
		Fort Union (Eocene).				Intermediate (Miocene).				Lamar Flora (Miocene).							
Species	Yellowstone River, half a mile below mouth of Elk Creek.	Yellowstone River, 1 mile below Elk Creek, opposite Hellroaring Creek.	Northeast side Crescent Hill.	Crescent Hill above mouth of Yancey's Creek.	Peak west of Passage, on Carnelian Creek.	Tower Creek.	Fossil Forest and ravine, 1 mile south of Yancey's.	Hill south of Yancey's.	Hill above mouth of Lost Creek.	Specimen Ridge west of Crystal Creek.	Fossil Forest a.	Cliff west of Fossil Forest, near Chalcedony Creek.	Lamar River, between Carbon and Calfee creeks.	Laramie coal deposits.	Denver and Livingstone.	Fort Union group.	Green River group.	Auriferous gravels California.
	1.	2.	3.	4.	5.	6.	7.	8.	9.	10.	11.	12.	13.	14.	15.	16.	17.	18.
69. Dryophyllum longepetiolatum n. sp.												7						
70. Ulmus pseudo-fulva? Lx																	×	
71. Ulmus minima? Ward																	×	
72. Ulmus rhamnifolia? Ward																		
73. Ulmus, fruits of																		
74. Planera longifolia Lx											3, 5					×	
75. Ficus deformata n. sp																×	
76. Ficus ungeri Lx																	×	
77. Ficus sp																		
78. Ficus elastica-nea? Lx																		
79. Ficus sordida Lx											6							
80. Ficus densifolia n. sp											3							
81. Ficus laurgeri n. sp										M 8								
82. Ficus tiliæfolia? Al. Br																		
83. Ficus asimmetrica Lx											3							
84. Artocarpus? quercoides n. sp																		
85. Magnolia californica Lx																		
86. Magnolia spectabilis n. sp										M 8, 3								
87. Magnolia nervosophylla n. sp											3							
88. Magnolia culveri n. sp																		
89. Magnolia? pollardi n. sp																		
90. Laurus primigenia? Ung												7						
91. Laurus perlita n. sp																		
92. Laurus montana n. sp																		
93. Laurus princeps Heer																		
94. Laurus californica Lx											3, 5, 6							
95. Laurus grandis Lx											3, 5, 7							
96. Persea pseudo-carolinensis Lx																		
97. Malapoenna lanarensis n. sp																		
98. Malapoenna coneata n. sp																		
99. Cinnamomum spectabilis Heer																		
100. Platanus guilielmæ Göpp											3, 5, 6, 7			D. L. ×	×	×	

a The numbers refer to the beds in which the plants were found.

Table showing the distribution of the Tertiary plants of the Yellowstone National Park—Continued.

	Distribution in the Park.													Distribution outside.				
Species.	Fort Union (Eocene).					Intermediate (Miocene).			Lower Flora (Miocene).									
	Yellowstone River, half a mile below mouth of Elk Creek.	Yellowstone River, 1 mile below Elk Creek, opposite Hellroaring Creek.	Northeast side Crescent Hill.	Cinnabar Hill, above Gardi Valley.	Peak west of Hunters, on Carnelian Creek.	Tower Creek.	Fossil Forest and Amethyst, 1 mile south of Yancey.	Hill south of Yancey.	Hill above mouth of Lost Creek, or Specimen Ridge, near Crystal Creek.	Fossil Forest a	Bit west of Fossil Forest, near fossil redwood creek.	Lamar River, between Cooke and Yellowstone parks.	Lamar-coal beddings.	Deep and Livingston.	Fort Union group.	Green River group.	Auriferous gravels, California.
	1	2	4.	5.	6.	7.	8.	9.	10.	11.	12.	13.	14	15.	16	17	18
101. Platanus montana n. sp																	
102. Acacia mucronata n. sp								7									...
103. Acacia lamarensis n. sp																	...
104. Acacia wardi n. sp							4										...
105. Leguminosites cassidis Lx																	...
106. Leguminosites lamarensis n. sp.																	
107. Rhus mixta Lx																	
108. Celastrus alverii n. sp																	
109. Celastrus inaequalis n. sp																	
110. Celastrus ellipticus n. sp																	
111. Elaeodendron polymorphum Ward										3							
112. Acer vicarium n. sp										7							
113. Acer fruit of																	
114. Sapindus affinis Newby																	
115. Sapindus alatus? Ward																	
116. Sapindus grandifoliolus Ward b.										6							...
117. Sapindus grandifolioides n. sp						1											...
118. Sapindus wardi n. sp																	...
119. Rhamnus rectinervis Lx										3,7			D.L				...
120. Paliurus colombi Heer																	x
121. Zizyphus serrulatus Ward															x		...
122. Cissus haguei n. sp										M. 8							...
123. Pterospermites haguei n. sp																	...
124. Grohersia? pachyphylla n. sp																	...
125. Tilia populifolia Lx																	...
126. Grewiopsis? alderseni n. sp																	...
127. Aralia wrighti n. sp																	
128. Aralia notata Lx							4		7					x			x
129. Aralia serrulata n. sp																	
130. Aralia whitneyi Lx										4,7							x
131. Aralia sp																	
132. Cornus newberryi Hollick																	
133. Cornus wrighti n. sp																	...

a The numbers refer to the beds in which the plants were found   b Also on the Thunderer

Table showing the distribution of the Tertiary plants of the Yellowstone National Park — Continued.

	Distribution in the Park												Distribution outside				
Species	Fort Union Flora						Lamar Late (Miocene)			Lamar Flora (Miocene)								
134. Arctostaphylos elliptica n. sp.																		
135. Diospyros brachysepala A. Br.											●							
136. Diospyros lancaterensis n. sp.																		
137. Diospyros lingulata n. sp.																		
138. Fraxinus wrightii n. sp.																		
139. Phyllites crassatulus n. sp.																		
140. Carpites newberryi n. sp.																		
141. Carpites pedunculatus n. sp.																		
142. Sequoia magnifica n. sp.?																		
143. Phyoxylon abietoides n. sp.																		
144. Phyoxylon anetbysimum n. sp.																		
145. Laurinoxylon pulchellum n. sp.																		
146. Persoxylon aromaticum Felix																		
147. Platanoxylon haydeni Felix																		
148. Rhamnacinium radiatum Felix																		
149. Quercinium lamarense n. sp.																		
150. Quercinium knowltoni Felix																		

GEOLOGICAL CONSIDERATION OF THE TERTIARY FLORA.

Naturally the geological aspects of this fossil flora are considered as of paramount importance, for it was to ascertain the bearing of the plants on the question of geological age that this investigation was undertaken. As I have already pointed out under the section devoted to the biological aspects of the flora (p. 775), a very large proportion of the plants were found to be new to science, and therefore could have only limited value in determining the age, but enough previously described forms were recognized to warrant certain deductions. It is the purpose to set these conclusions forth in this section.

The first plants brought back from this portion of the Yellowstone National Park by the early Hayden survey parties were submitted to Prof. Leo Lesquereux, and although few in number the specimens and species were nevertheless sufficient to afford some indication of their age. Professor Lesquereux regarded the plants from Elk Creek and vicinity as indicating an Eocene age, and those from the well-known Fossil Forest on the west side of the Lamar Valley as closely allied to those of the Auriferous gravels of California. It is a pleasure to state that this adumbration has been abundantly confirmed by the results of more searching study of a larger amount of material, but at the time this was outlined the facts were so few that the suggestions were not regarded as conclusions, and as it was before any careful detailed geological work had been done, these now clearly defined horizons came to be grouped together under the somewhat noncommittal term Volcanic Tertiary.

Although the geology of the region has been fully discussed by Mr. Arnold Hague in Part I of this monograph, it seems necessary, for the satisfactory understanding of the problems requiring solution, to set forth briefly the geological features characterizing this portion of the Park which have a direct bearing upon the remarkable flora found.

In the first place, all the material constituting the beds in which the Tertiary plants are embedded is of volcanic origin. According to the geologists, this material may be divided into two distinct periods of volcanic eruption, which may be distinguished by their mineral composition.

The older series of these lavas has been designated as the early acid breccias and flows, and the younger as the early basic breccias and flows. Both these series of rocks carry plant remains. In general the matrix in which the plants are preserved is a fine-grained ash, probably deposited as a mud flow, with all the appearance of stratification and other indications of water-laid deposits. Occasionally the material is much coarser and has the appearance of breccias mixed with fine sediments. The acid rocks are usually light in color—yellow, lavender, or gray—while the basic rocks, which carry more iron, are darker in color—frequently some shade of green or dark brown, passing over into black. In some instances, as might be expected with fine water-laid beds, the deposits in both series of lavas closely resemble one another, while the great mass of lava of the two bodies may be readily distinguished. The acid breccias, the oldest of the lava flows, rest in many places upon the upturned edges of Archean and Paleozoic rocks. In most instances the basic breccias either rest directly upon the acid rocks or else the underlying rocks are not exposed. The acid breccias are found on both banks of the Yellowstone River near the mouth of Elk Creek, and near the junction of the Yellowstone River with Hellroaring Creek, as well as on Crescent Hill and near the head of Tower Creek.

In the neighborhood of Lost Creek, and on the northern end of Specimen Ridge, along the drainage of Crystal Creek, the basic breccias are known to lie directly upon the acid breccias. In these localities the flora has a character distinctly its own, and bears evidence of being younger than the flora from the acid breccias. The basic series of rocks is typified at the Fossil Forest, and also at the cliff a short distance to the south and east of the Fossil Forest. They occur also on the east bank of Lamar River, between Cache and Calfee creeks. All of these localities are characterized by their plant remains, and the following determinations of age are fully warranted.

The table of distribution of Tertiary plants in the Yellowstone National Park has been prepared for the purpose of showing in a graphic manner the distribution of the various plants within the limits of the Park and the affinities of those having an outside distribution. From this it appears that the Tertiary flora consists of 150 more or less satisfactory species. Of this number, 81 species, or over 50 per cent, are here described for the first time. New species can not, of course, have the value in determining age that previously described forms have, but when their general facies as well as close affinities are taken into account, they also become of positive value. On eliminating the 81 new species, together with 8 forms not specifically named, there remain 61 species upon which we must depend in the determination of the ages of the various strata in which they are contained.

A further examination of the table brings out the fact that this flora may be naturally divided into 3 more or less distinct subfloras or stages. These three divisions are the older or acid series, the intermediate series, and the basic or younger series. The first division (acid) has a flora of 79 species; the second (intermediate) a flora of 30 species, and the third (basic) a flora of 70 species. It further appears that 23 species or forms are common to two or more of the divisions.

The flora of the older or acid series will be first considered. Of the 79 species, 42 are either new to science or not specifically named, leaving 37 species having a distribution beyond the limits of the Park. Following is a list of these species:

Asplenium erosum Lx.;
Lygodium kaulfusii (Heer).
Taxites olriki Heer.
Sequoia couttsiæ Heer.
Sequoia langsdorfii (Brgt.) Heer.
Sparganium stygium Heer.
Cyperacites angustior Al. Br.
Musophyllum complicatum Lx.
Juglans rugosa Lx.
Juglans schimperi Lx.
Myrica scottii Lx.
Populus glandulifera Heer.
Populus speciosa Ward.
Populus daphnogenoides Ward.
Salix lavateri Heer.
Fagus antipofii Heer.
Quercus consimilis Newby.
Quercus olafseni Heer.

Ulmus rhamnifolia Ward.
Ficus ungeri Lx.
Ficus asiminæfolia Lx.
Laurus primigenia ? Ung.
Laurus princeps Lx.
Laurus californica Lx.
Laurus grandis Lx.
Cinnamomum spectabilis Heer.
Platanus guillelmæ Göpp.
Leguminosites cassioides Lx.
Sapindus affinis Lx.
Sapindus alatus Ward.
Sapindus grandifolius Ward.
Paliurus colombi Heer.
Zizyphus serrulatus Ward.
Tilia populifolia Lx.
Aralia notata Lx.
Cornus newberryi Hollick.

These 37 species have the following distribution: Five are found in the coal-bearing Laramie, 5 in the Denver and Livingston, 17, or nearly 50 per cent, in the Fort Union, 9 in the Green River group, and 11 in the Auriferous gravels of California. Of the species common to the acid rocks and the Laramie at Denver and Livingston beds, not one is found exclusively in these beds, but they are such species as *Sequoia langsdorfii, Juglans rugosa, Platanus guillelmæ,* and *Juglans schimperi,* which enjoy a wide geological and geographical distribution.

The Fort Union element in this flora is a very important one; in fact, it may be called the dominant element. It includes at least 12 species that have never before been found outside of the type locality. Among these are *Sparganium stygium, Populus speciosa, Populus daphnogenoides, Ulmus minima, Ulmus rhamnifolia, Sapindus affinis, Sapindus grandifolidus,* and *Cornus newberryi.* Some of these are represented by as many as 200 individuals, showing that they existed in great abundance, as they are also known to have existed at the mouth of the Yellowstone. This abundance

also makes their determination certain. Several other species, having a
wider distribution, are very abundant in these beds, such as *Aralia notata*,
which is represented by more than 100 specimens, and *Sequoia langsdorfii*,
which has a wide distribution, but is most abundant in this country in
the Fort Union beds. *Sequoia couttsiae*, having a somewhat wide range, is
also very abundant in the beds under consideration and the Fort Union.
Besides these are a number of species that can not be mistaken, as *Zizyphus
serrulatus*, *Taxites olriki*, etc.

One species, *Asplenium erosum*, has been found in both Laramie and
Denver strata in Colorado. It is represented by only 2 or 3 small and
more or less doubtful examples from the Yellowstone below Elk Creek.
Juglans rugosa is a species of wide distribution and therefore of comparatively little value stratigraphically. It is found from the Laramie to the
Miocene, but is rare in the acid beds within the Park. *Quercus olafseni* has
been found in the Laramie, but its determination in the Park depends on a
single doubtful fragment from the vicinity of Elk Creek.

The species that have also been found in the Green River beds are
comparatively unimportant. *Lygodium kaulfusii* is, in this country, a typical
Green River species. It is rare in the acid series, but abundant in the basic
series along the Lamar River. *Mosophyllum complicatum* has never before
been reported outside of the Green River beds. *Ficus ungeri* and *Tilia
populifolia* are typical Green River plants, but are represented here by one
or two examples each.

The species found in the Auriferous gravels are the only ones remaining
to be considered. Of the 11 species, *Juglans rugosa*, *Quercus breweri*, *Salix
lavateri*, and *Quercus olafseni* are open to doubt, as they are represented by
only one or two fragments each. *Ficus asiminiaefolia* likewise depends upon
a single leaf, but it is a well-preserved one, and the determination is probably
correct. *Aralia notata*, another of the species, is very rare, if really found
at all, in the Auriferous gravels. The three remaining species are relatively abundant, and there is little question as to the correctness of their
determination.

The species whose distribution lies beyond the limits of the Park
having been passed in review, it will be of interest to note the obvious
affinities of certain of the more important new forms. Thus, *Asplenium
remotidens* is closely related to *A. erosum*, and *Dryopteris woodii* and *D. xantho-*

lithosis to *Lastrea goldiauum*, both of which are abundant in the Denver beds of Colorado. *Juglans crescentia* is related to *J. nigella*, as identified by Professor Ward in the Fort Union group. The beautiful new *Populus cuullolithensis* is very close to *P. genatric* of Newberry, from the Fort Union group. *Betula iddingsi* is obviously related to *B. prisca*; *Quercus gomcqi* to *Q. laurifolia*; *Platanus montana* to *P. raynoldsii*; *Celastrus culveri* to *C. ovatus* and *C. curvinervis*; and *Sapindus grandifolioloides* to *S. grandifoliolus*, all of the Fort Union group.

From this evidence it appears that the flora of the early acid breccias in the Yellowstone National Park finds its closest affinity with the flora of the Fort Union group, and it is unhesitatingly referred to that age. The relation of this flora to that of the Laramie is unimportant, being confined to less than half a dozen species. Its relationship to the Denver and Green River floras is naturally closer, but it forms only a small element of these, as also with the flora of the Auriferous gravels of California. The relation, as based on total number of species, is unimportant, but in the upper beds it begins to show a transition.

It will be next in order to consider the intermediate flora. As already stated, this embraces 30 species, of which number 16 are confined to these beds and 14 are distributed outside, either in the acid or basic series or beyond the limits of the Park. A further analysis brings out the fact that of the 16 species peculiar to these beds 13 are regarded as being new to science, and of the 14 species found beyond the limits of these beds 5 are new to science. This makes a total of 18 species that are regarded as new, leaving 12 species having a distribution without the Park. Following is a complete list of these 12 species:

Osmunda affinis Lx.	Laurus grandis Lx.
Sequoia langsdorfii Brgt.) Heer.	Platanus guillelmae Göpp.
Juglans rugosa Lx.	Elaeodendron polymorphum Ward.
Uhous minima Ward.	Sapindus affinis Lx.
Ficus tiliaefolia Al. Br.	Aralia notata Lx.
Laurus californica Lx.	Aralia whitneyi Lx.

Four of the species above enumerated (*Sequoia langsdorfii*, *Juglans rugosa*, *Platanus guillelma*, and *Aralia notata*) have a wide distribution, being found from the Laramie to the Upper Miocene, and are therefore of comparatively little value. One of the remaining (*Osmunda affinis*) is found in the Denver beds; 3 are confined to the Fort Union, and 6 species are found in the Auriferous gravels of California.

Of the 5 new species found in other beds, 2 are common to the older or acid series and 3 to the younger or basic series. It therefore becomes apparent that this flora, although reasonably distinct, finds its greatest affinity with the younger or basic series. This is shown by the species common to the intermediate series and the Auriferous gravels and by the new species common to the basic series. This is not especially pronounced, and could hardly be made out for the intermediate flora as a whole, were it not for certain species that come from rocks that are directly succeeded by the basic rocks. For these reasons it was at first supposed that a part of the localities represented belonged to the lower and a part to the upper beds, but by combining several of these this intermediate flora was worked out. But, as stated above, the rocks of this series that are known to be the lowest bear a flora nearest to that of the acid rocks, and the rocks known to be higher hold plants most nearly related to those of the upper or younger beds.

It now remains to consider the flora of the basic breccias and its relationships. The typical locality for this flora is the Fossil Forest and vicinity, including the locality on the east side of the Lamar River, between Cache and Calfee creeks. This flora, as a whole, embraces 70 species or forms, distributed as follows: 38 species new to science, 3 forms not specifically named, and 29 species having a distribution beyond the limits of the Park. Following is a list of the species having an outside distribution:

Lygodium kaulfusii Heer.
Sequoia langsdorfii (Brgt.) Heer.
Juglans californica Lx.
Juglans rugosa Lx.
Hicoria antiqua (Newby.) Kn.
Populus balsamoides Gœpp.
Salix varians Heer.
Salix angusta Al. Br.
Salix elongata Heer.
Corylus macquarryi (Forbes) Heer.
Quercus furcinervis americana Kn.
Ulmus pseudo-fulva Lx.
Planera longifolia Lx.
Ficus shastensis? Lx.

Ficus sordida Lx.
Ficus asiminiæfolia Lx.
Magnolia californica Lx.
Laurus californica Lx.
Laurus primigenia? Ung.
Laurus grandis Lx.
Persea pseudo-carolinensis Lx.
Platanus guillelmæ Gœpp.
Rhus mixta Lx.
Elæodendron polymorphum Ward.
Sapindus grandifolius Ward.
Rhamnus rectinervis Lx.
Aralia notata Lx.
Aralia whitneyi Lx.

Of these 29 species, only 4 have been found in the true Laramie. These are *Juglans rugosa*, *Platanus guillelmæ*, *Rhamnus rectinervis*, and *Diospyros brachysepala*, the last open to doubt. All of these species have a wide vertical range and are consequently of little value in indicating age. The affinities of this flora with that of the Laramie may therefore be regarded as unimportant.

The relationship of this flora with the Fort Union, Denver, and Green River groups is also relatively unimportant. There are 7 species found in each of these groups, but none are confined to the Denver, and only 1 to the Green River, and 3 to the Fort Union. The rest are of wide geographical and geological distribution.

The relationship of the flora of the basic rocks is clearly with that of the Auriferous gravels of California, for no fewer than 17 of the 29 species are common to the two localities, and 12 of the species are exclusively confined to them. These are such important species as *Aralia whitneyi*, *Persea pseudo-carolinensis*, *Laurus californica*, *Laurus grandis*, *Magnolia californica*, *Ficus sordula*, *Juglans californica*, *Rhus mista*, etc. Most of these are present in numbers in the Park flora, and there can therefore be no question as to the correctness of their identification.

Besides the species above enumerated that have actually been found common to the two localities, the numerous new species are in many cases unmistakably related to species known only from the Auriferous gravels. Thus, *Magnolia culveri* is close to *M. californica*, and *Magnolia spectabilis* is so close to *M. lanceolata* that Lesquereux so identified it. Other examples might be given, but they are unnecessary. The preponderance of evidence points to the similarity of age between the flora of the basic series and that of the Auriferous gravels of California. The fixing of the exact age of the Auriferous gravels is not a difficult matter. They were at first supposed to be Lower Pliocene in age, but the latest evidence, derived from a more or less complete restudy of the abundant flora, together with a thorough investigation of the stratigraphy, makes it reasonably certain that it is really Upper Miocene. This is the view taken in the present instance, and this flora in the Yellowstone National Park is referred to the Upper Miocene:

Asplenium iddingsii n. sp.	Magnolia ? pollardi n. sp.
Lygodium kaulfussii.	Laurus primigenia ? Ung.
Equisetum canaliculatum n. sp.	Laurus californica Lx.

Equisetum deciduum n. sp.
Sequoia langsdorfii (Brgt.) Heer.
Juglans rugosa Lx.
Juglans crescentia n. sp.
Castanea pulchella n. sp.
Ficus densifolia n. sp.
Ficus asiminaefolia Lx.

Laurus grandis Lx.
Litsea kamarensis n. sp.
Platanus guillelmae Göpp.
Elaeodendron polymorphum Ward.
Sapindus grandifoliolus Ward.
Sapindus wardii n. sp.
Aralia notata Lx.

Some of these species, as *Laurelium kaulfusii, Castanea pulchella, Laurus grandis, Platanus guillelmae, Sapindus grandifoliolus,* and *Aralia notata,* are important well-marked species that have weight in showing the close relationships between the floras of the two series; but, on the other hand, the perfect distinctness of the beds is shown by the fact that there are some 40 species that are confined to each horizon. It will not, therefore, be difficult in future to determine the horizons of the various plant-bearing beds within the Yellowstone National Park.

In order to show how remarkably distinct these three floras are, it will be necessary only to consider the species in common between them. As already stated, only 23 forms out of the total of 147 forms are common to two or more of the series of beds. It will not be necessary to present a complete list of these species in common. The numerical results show that 8 forms only are common to the three beds, 2 to the acid and intermediate, and 3 to the intermediate and basic, and, finally, that 10 are common to the acid and basic. When these facts are presented in connection with the total flora of each series, the differences become even more marked. Thus, the lower or acid series, with a flora of 79 species, has only 20 species common to the others. Of these, 8 are common to all three beds, 2 to it and the intermediate beds, and 10 to it and the upper or basic beds. The intermediate beds, with a flora of 30 species, have 13 species in common with the others. Of these, 8, as above stated, are common to all three, 2 to intermediate and acid, and 3 to intermediate and basic. The basic or younger beds, with a flora of 70 species, have 20 species common to the others. Of these, it is hardly necessary to repeat, 8 are common to all three, 3 to it and intermediate, and 10 to it and acid. These 3 floras are, therefore, shown to be markedly distinct, and it will not be difficult to distinguish them in future.

PLATE LXXVII.

PLATE LXXVII.

794

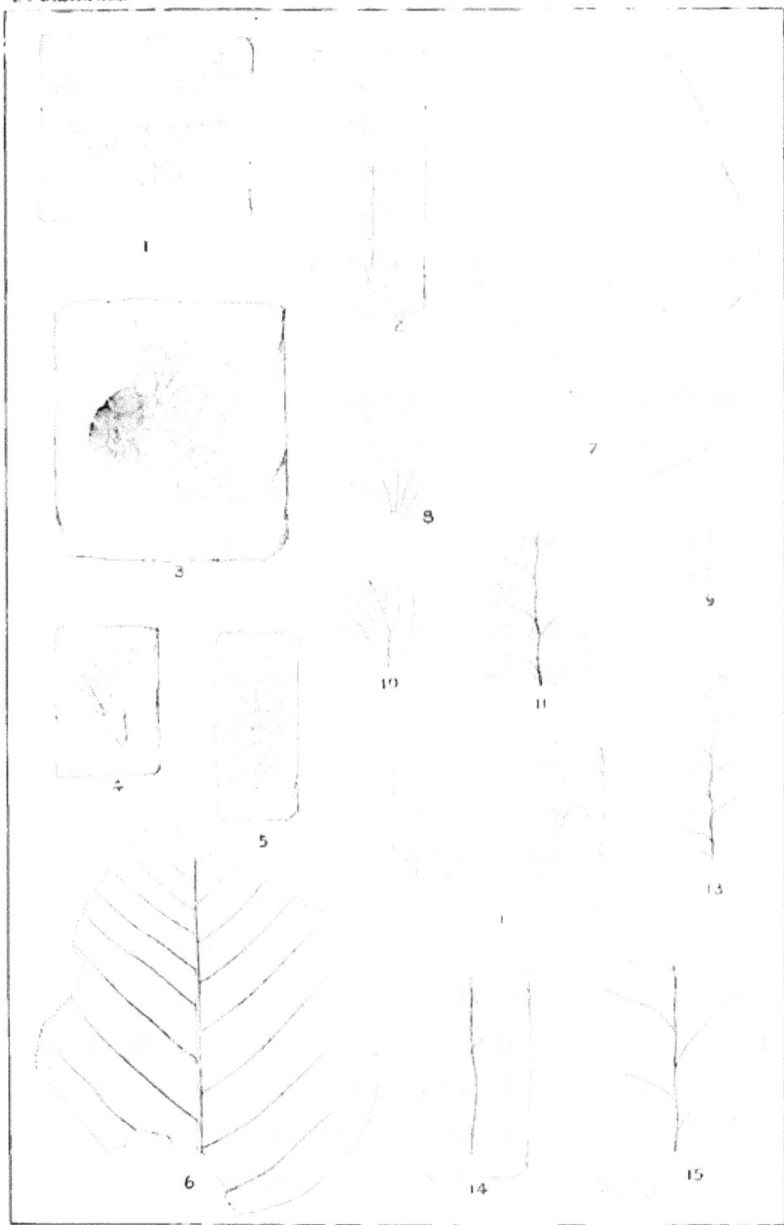

PLATE LXXVIII.

26

PLATE LXXVIII.

1

2

3

4

5

6

7

8

9

LARAMIE FORMATION

PLATE LXXIX.

PLATE LXXIX

798

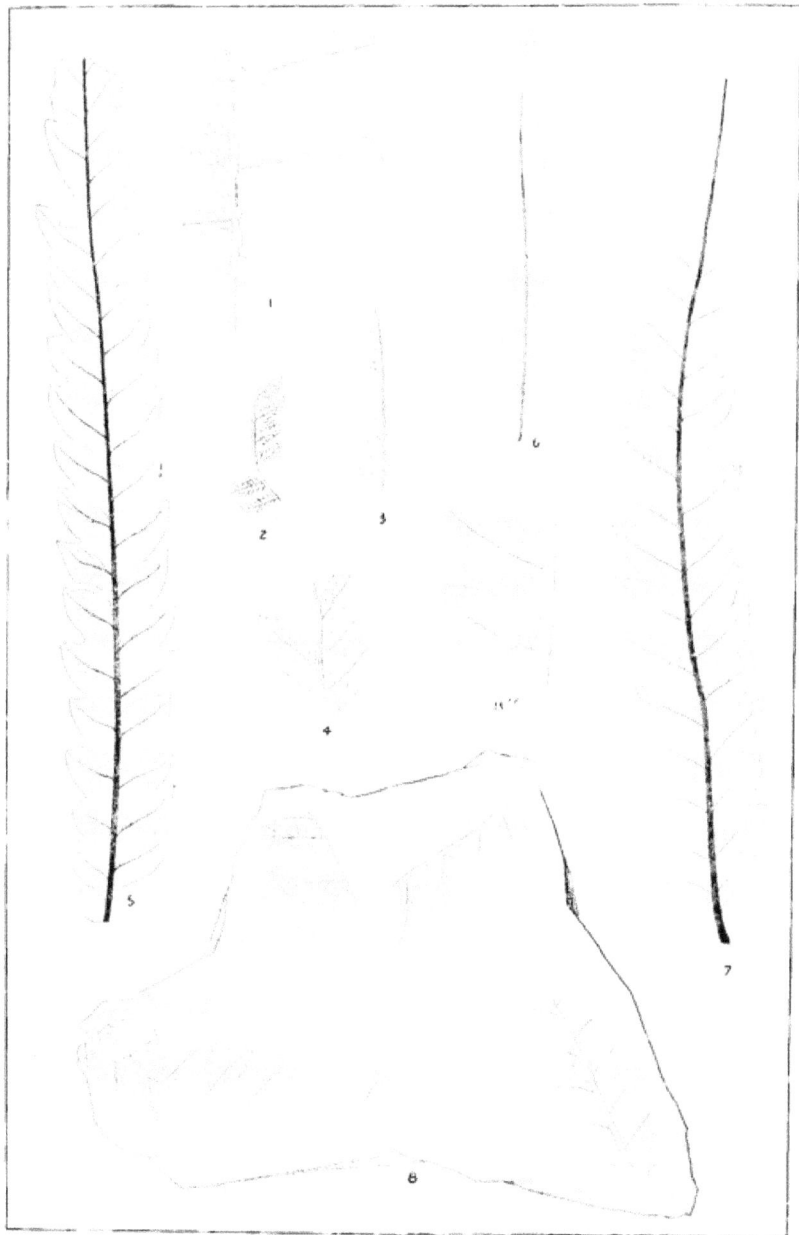

PLATE LXXX.

700

PLATE LXXX.

PLATE LXXXI.

PLATE LXXXI.

PLATE LXXXII.

PLATE LXXXII

804

PLATE LXXXIII.

PLATE LXXXIII.

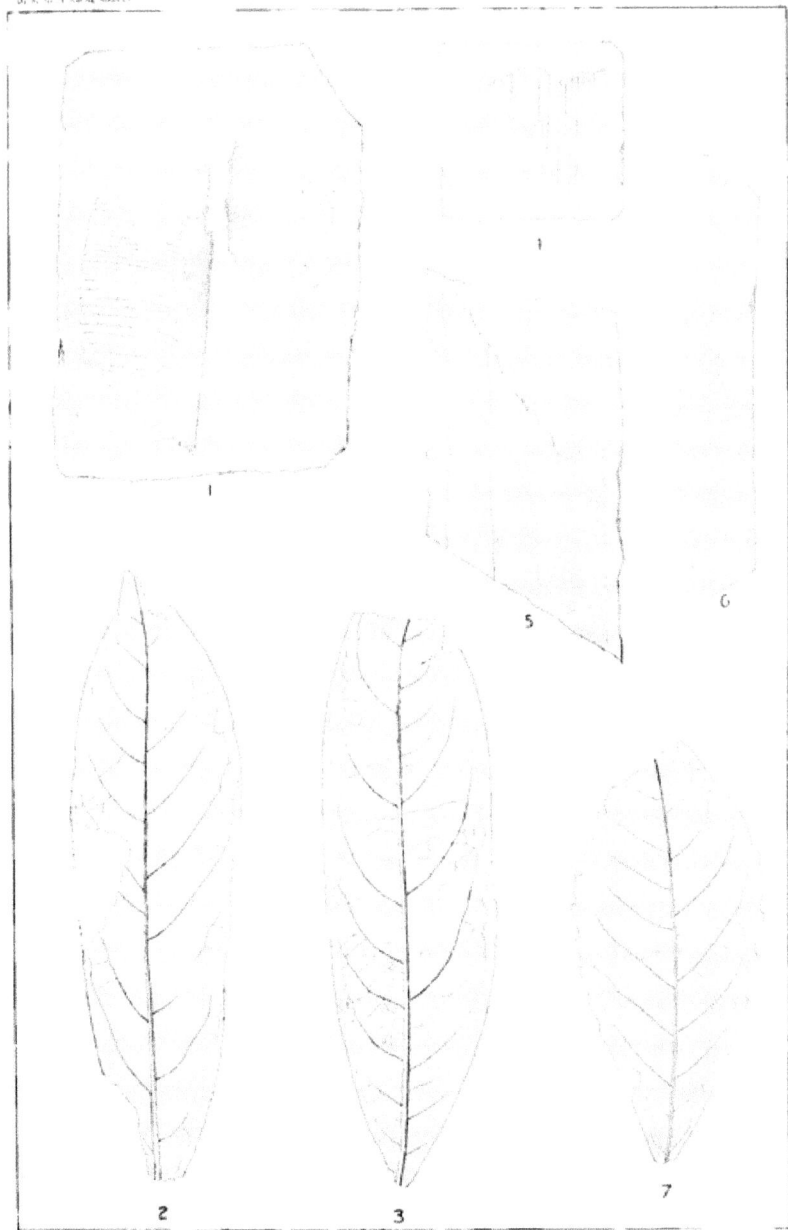

1

5

6

2

3

7

PLATE LXXXIV.

PLATE LXXXIV.

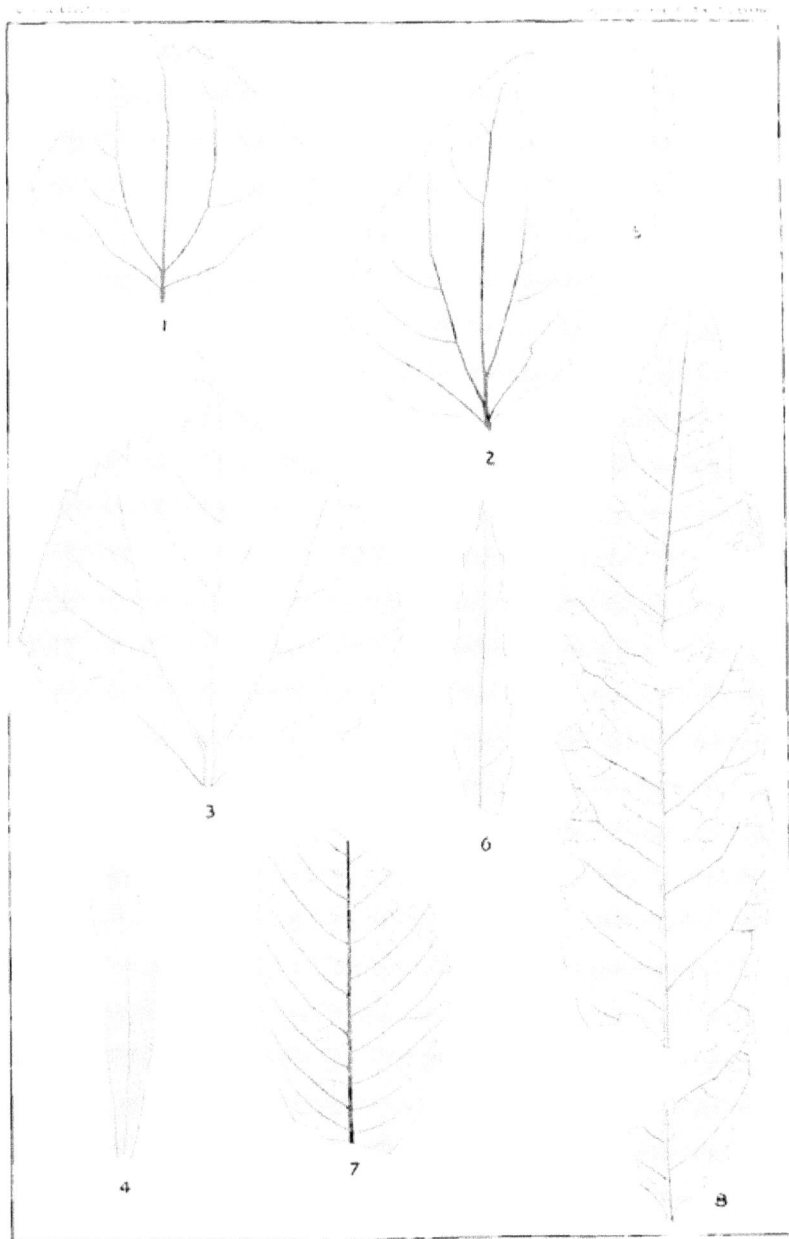

1

2

3

4

6

7

8

PLATE LXXXV

PLATE LXXXV.

810

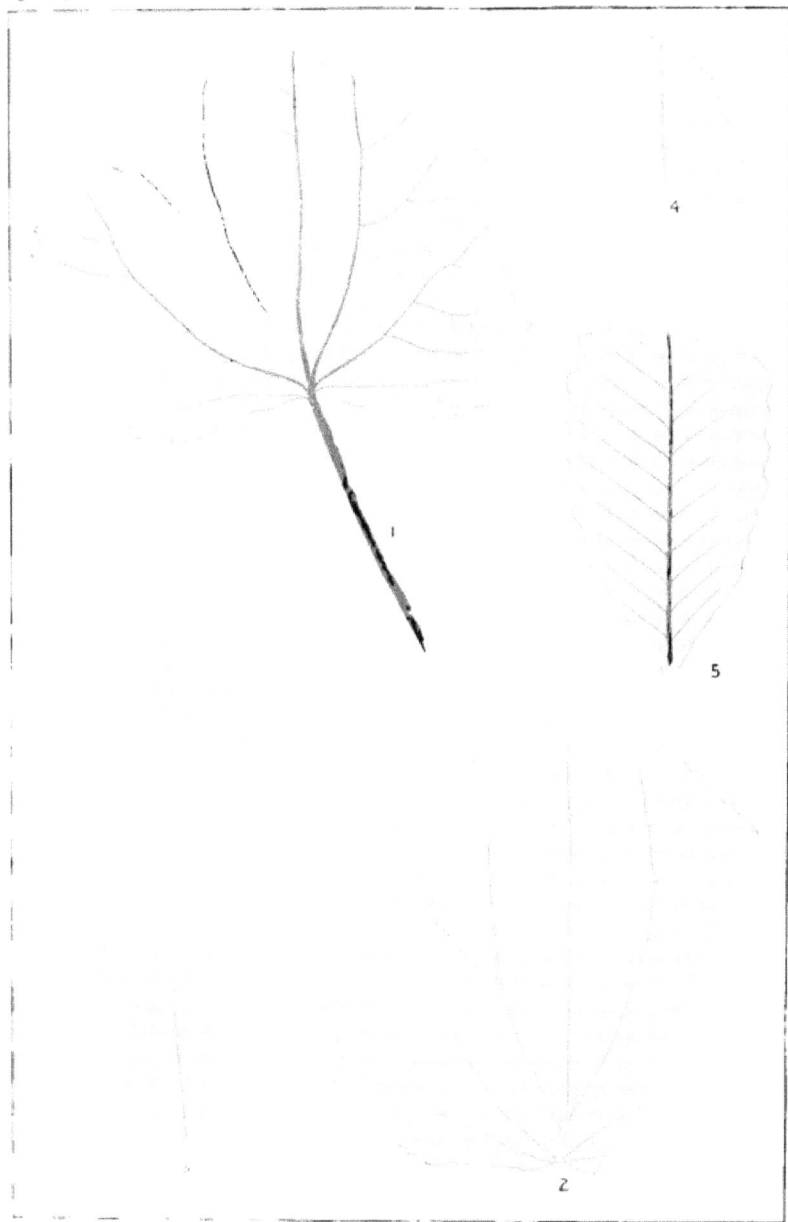

PLATE LXXXVI.

PLATE LXXXVI.

812

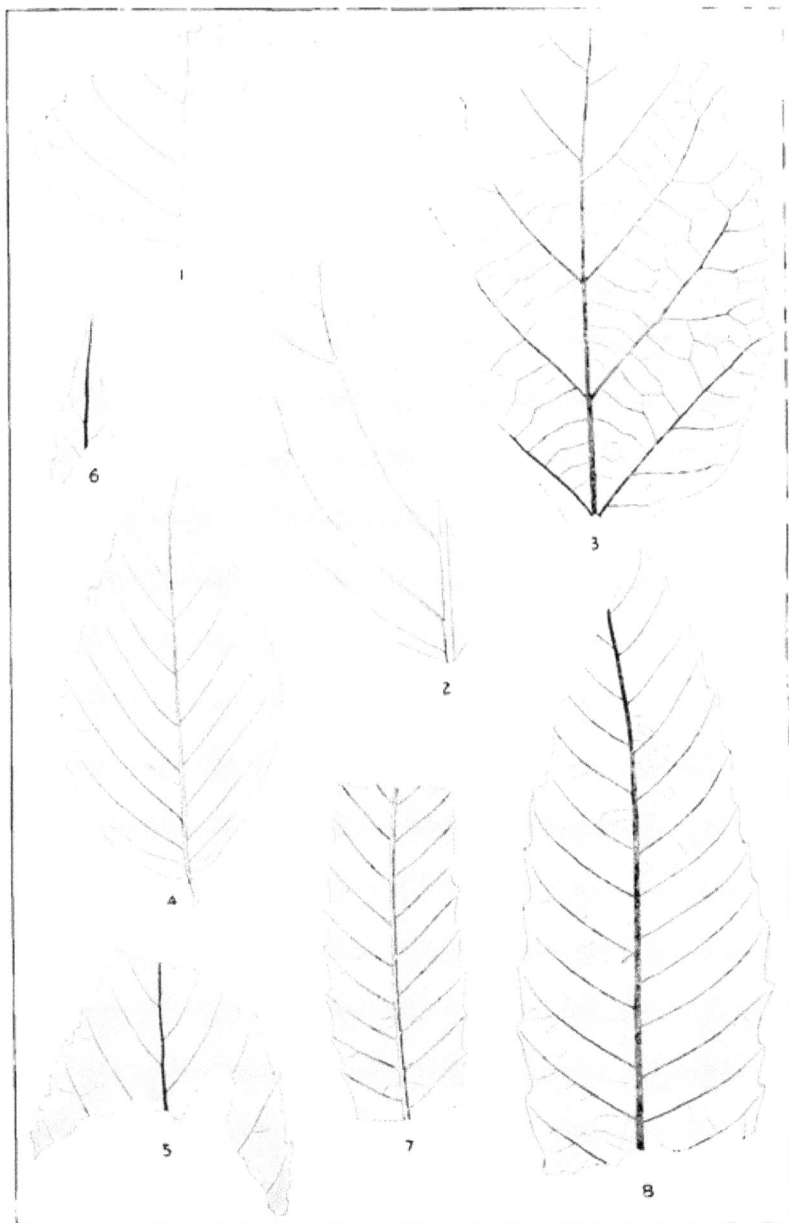

1

6

3

2

4

7

5

8

TERTIARY

PLATE LXXXVII.

PLATE LXXXVII.

PLATE LXXXVIII.

PLATE LXXXVIII.

PLATE LXXXIX.

PLATE LXXXIX

818

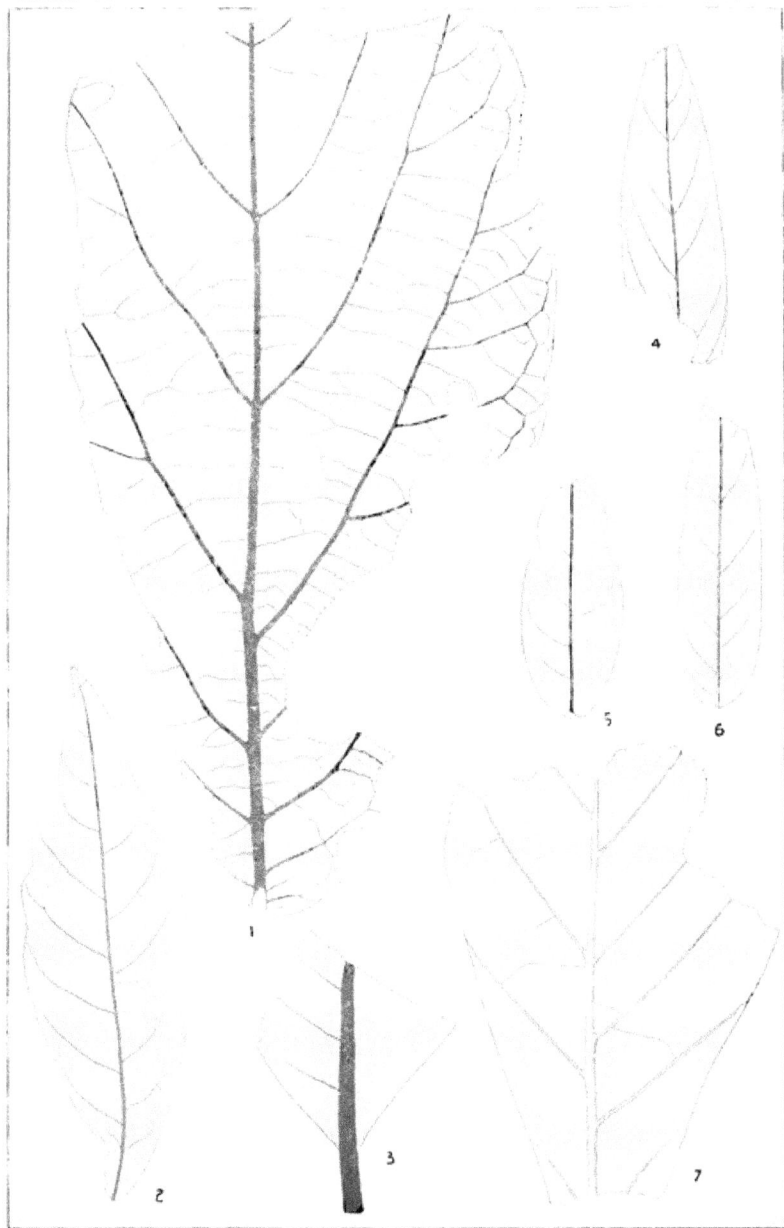

PLATE XC.

PLATE XC.

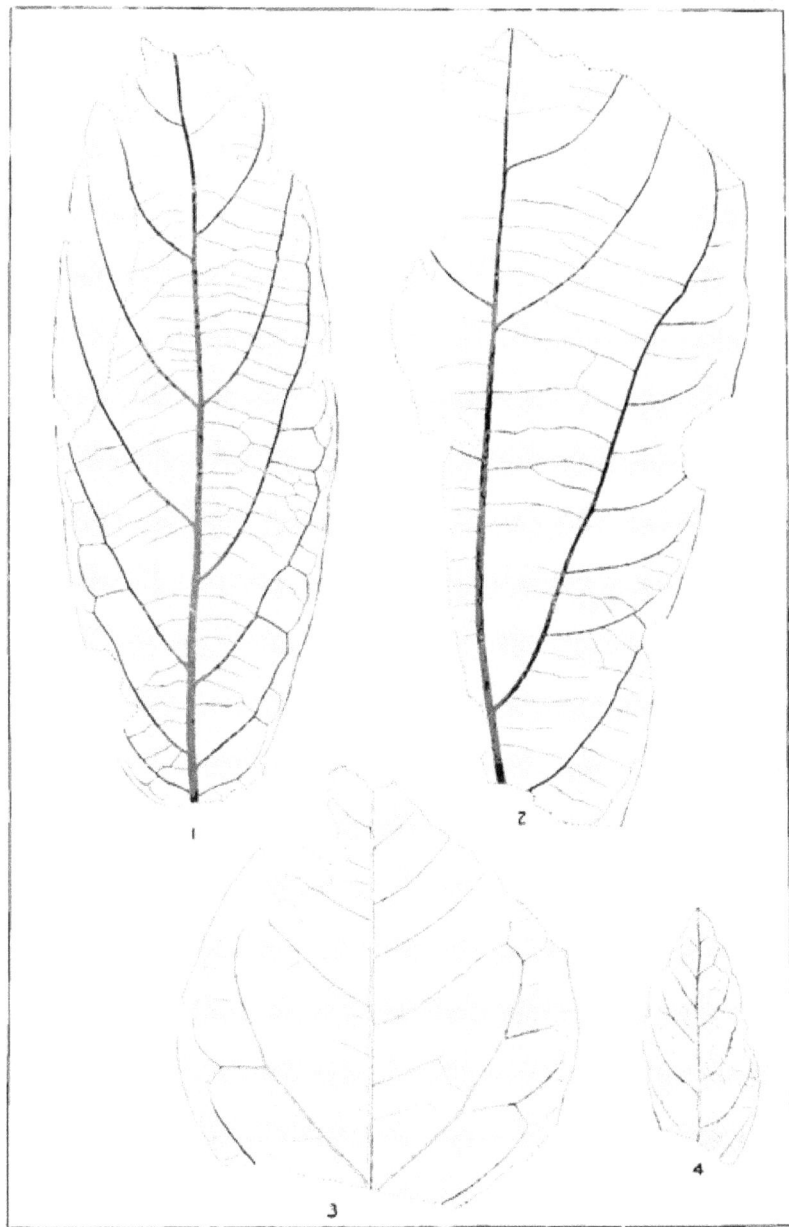

PLATE XCI.

PLATE XCI.

822

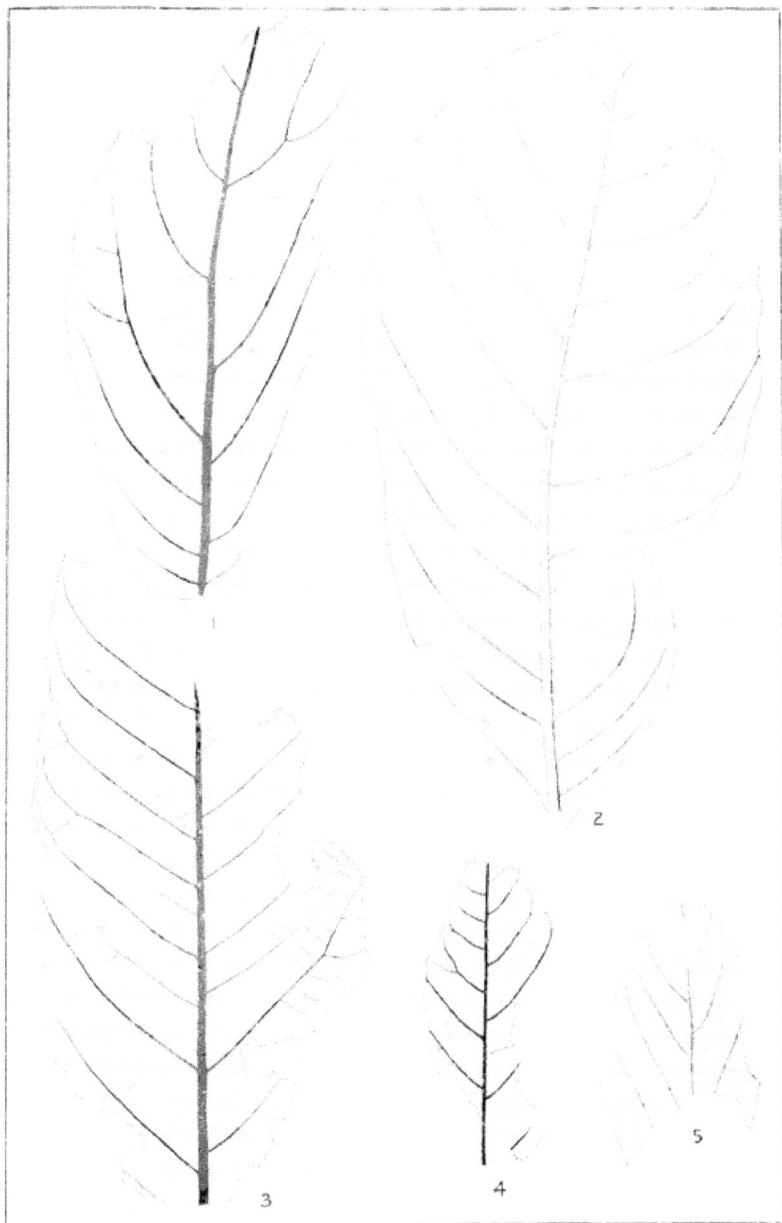

PLATE XCII.

823

PLATE XCII.

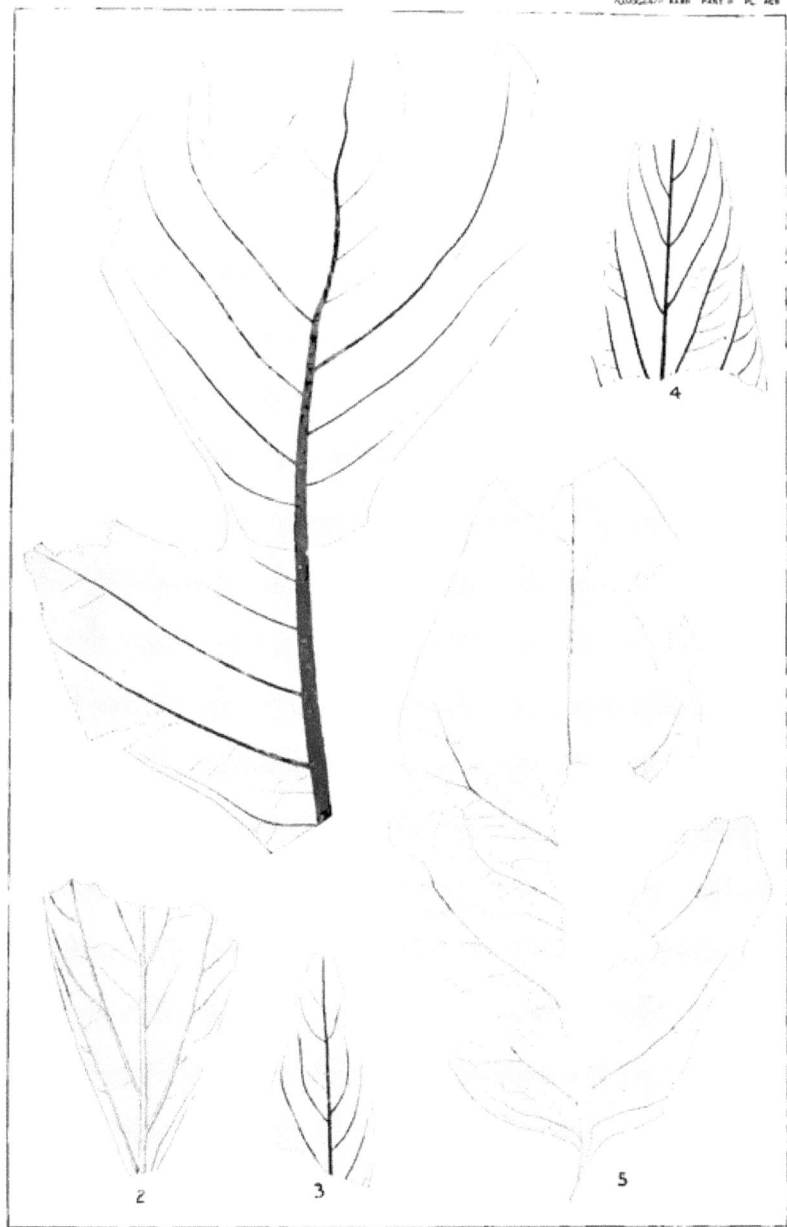

2   3   4   5

PLATE XCIII.

PLATE XCIII.

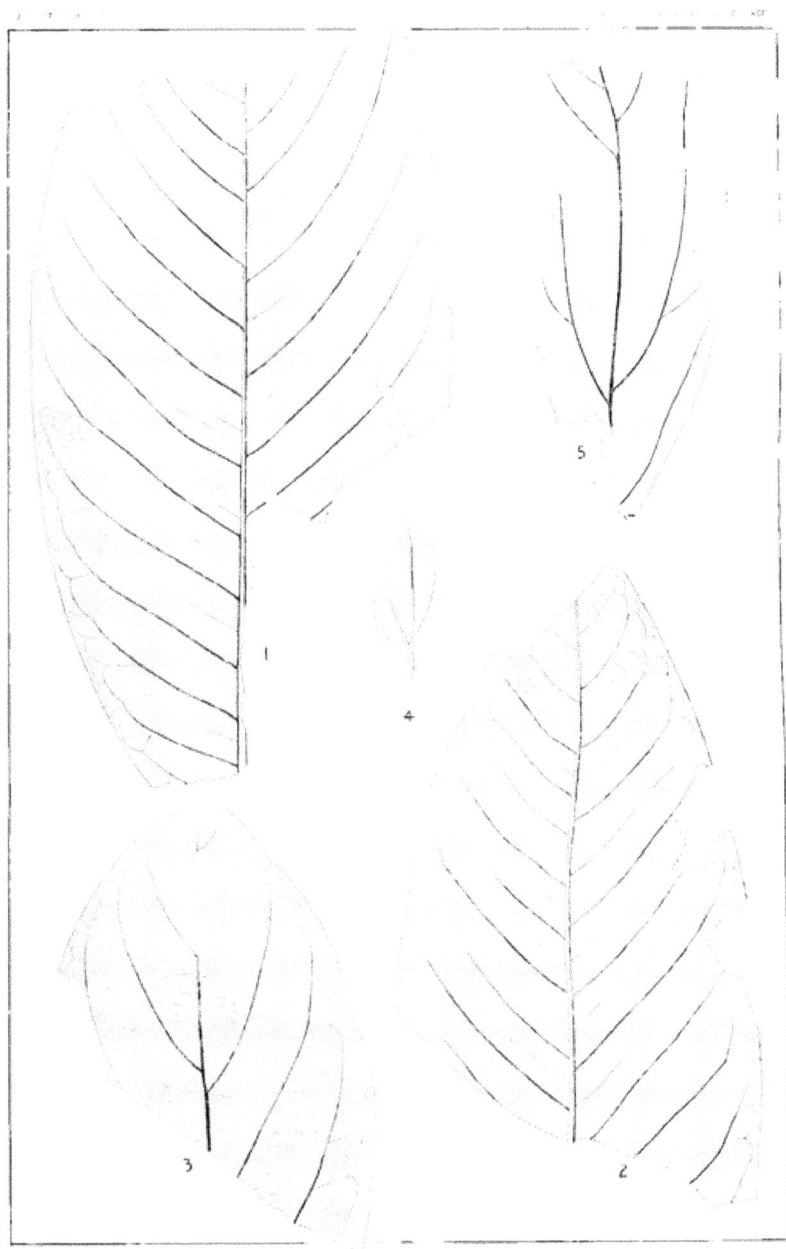

PLATE XCIV.

PLATE XCIV.

828

2

3

4

5

6

PLATE XCV.

PLATE XCV.

850

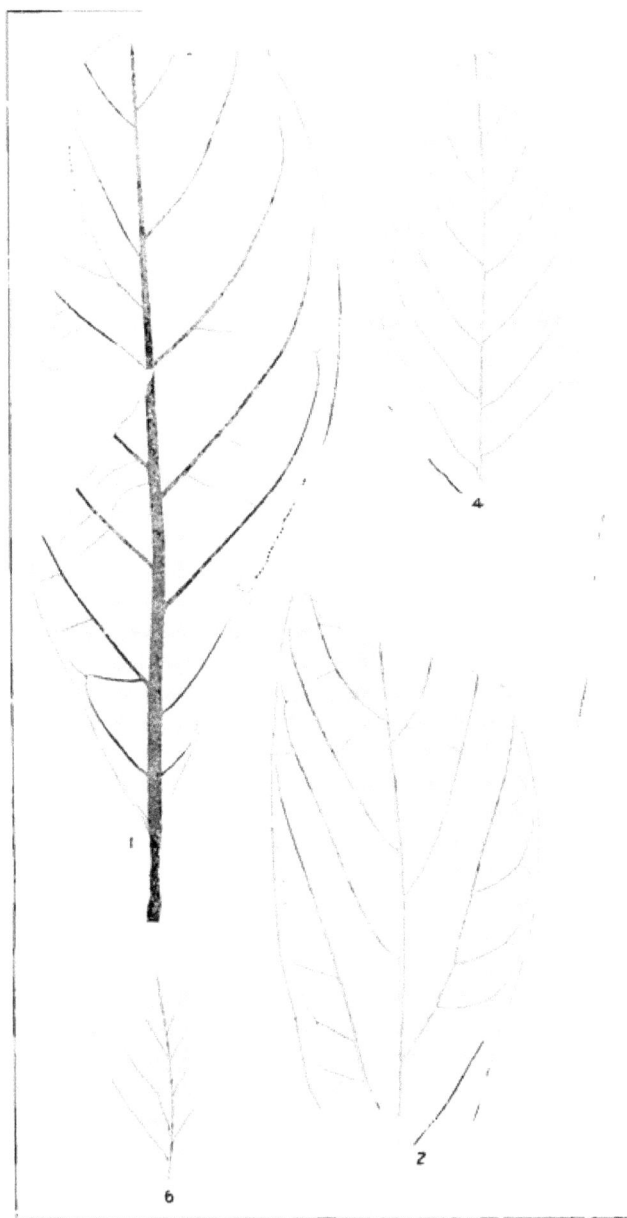

PLATE XCVI.

PLATE XCVI.

882

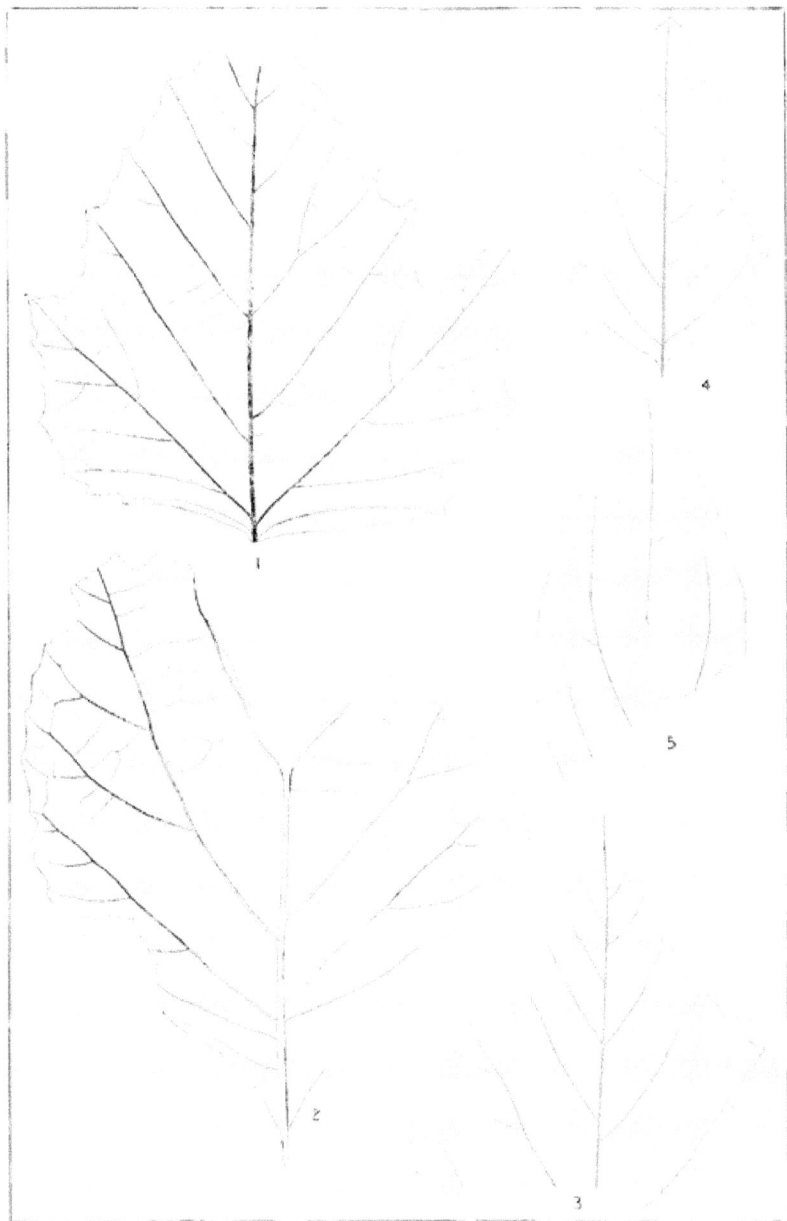

1

2

3

4

5

TERTIARY

PLATE XCVII.

PLATE XCVII.

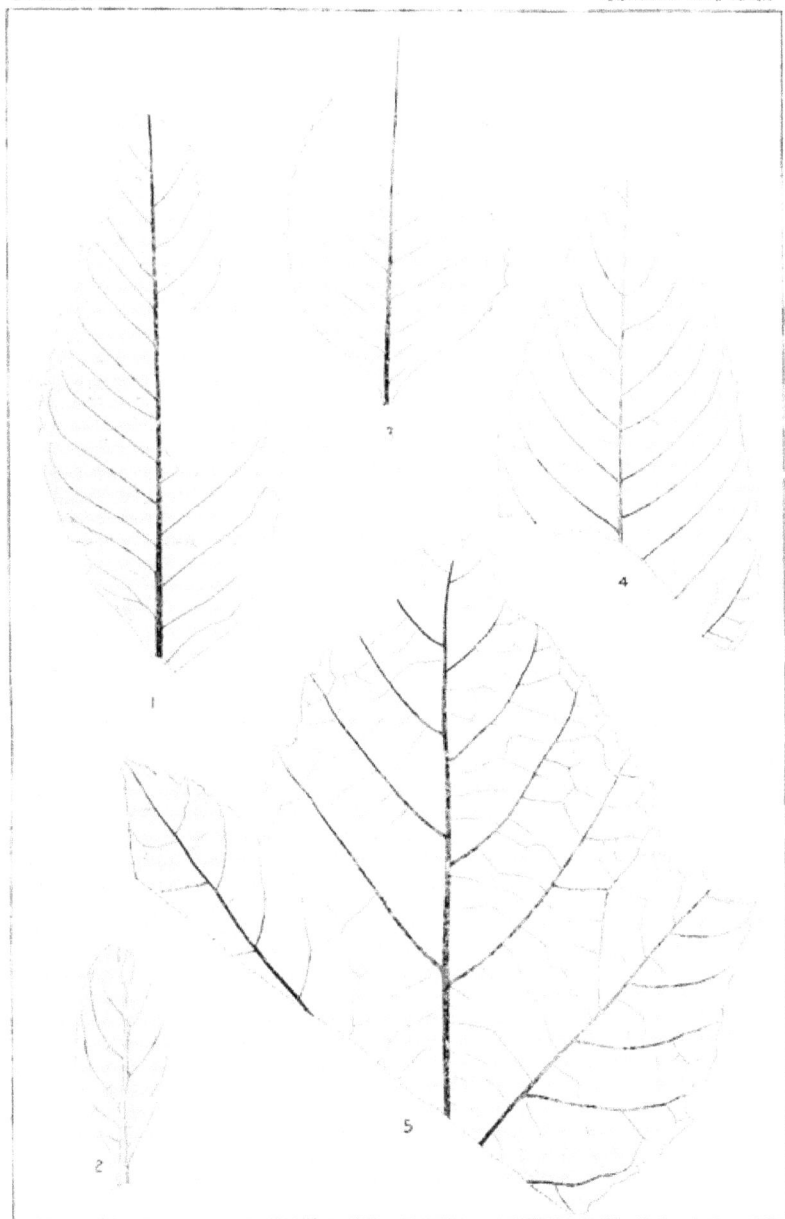

PLATE XCVIII.

PLATE XCVIII.

PLATE XXI

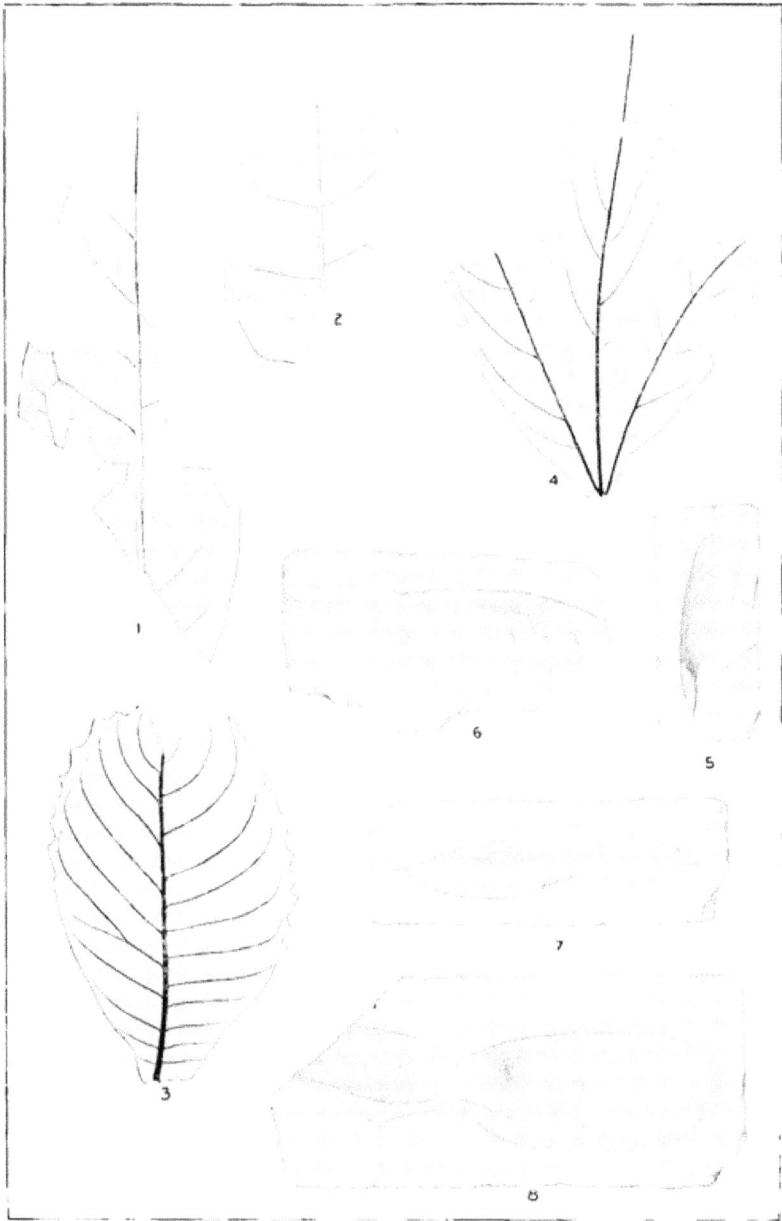

PLATE XCIX.

PLATE XCIX.

838

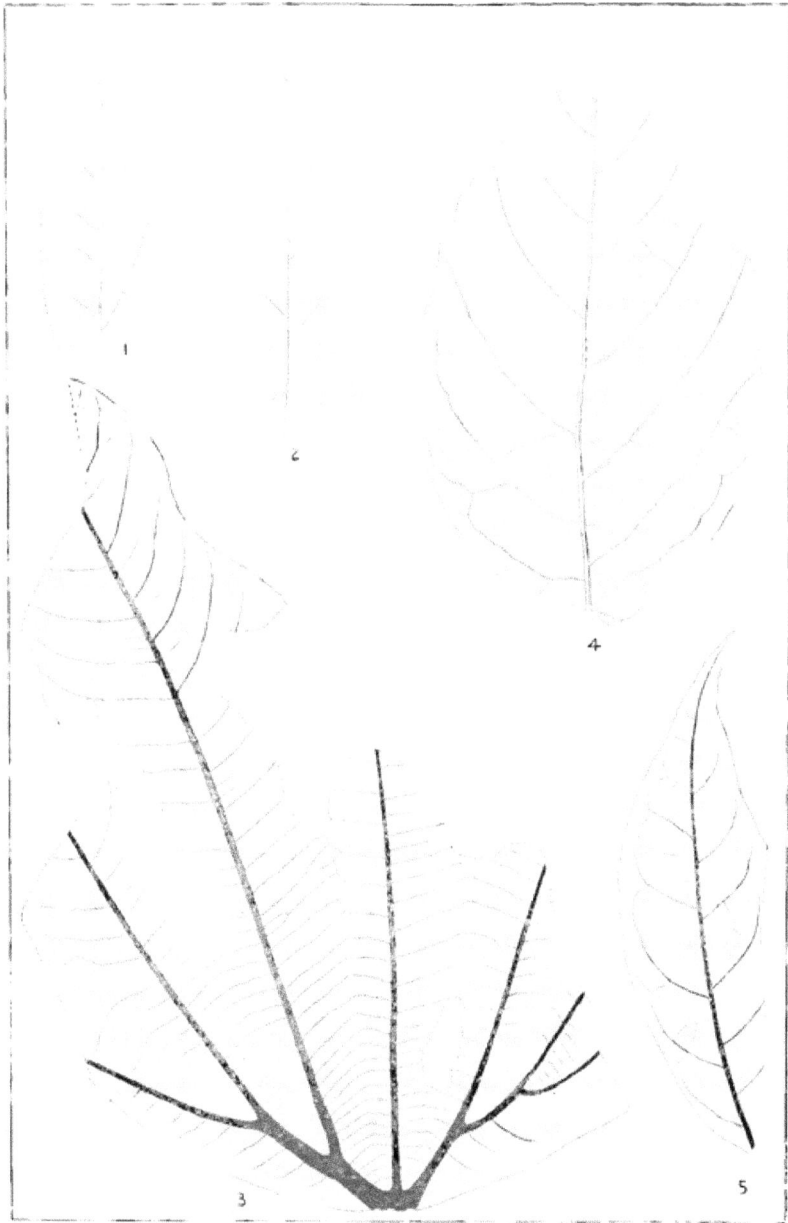

TERTIARY

PLATE C.

PLATE CL.

PLATE CI.

842

PLATE CII.

PLATE CII.

844

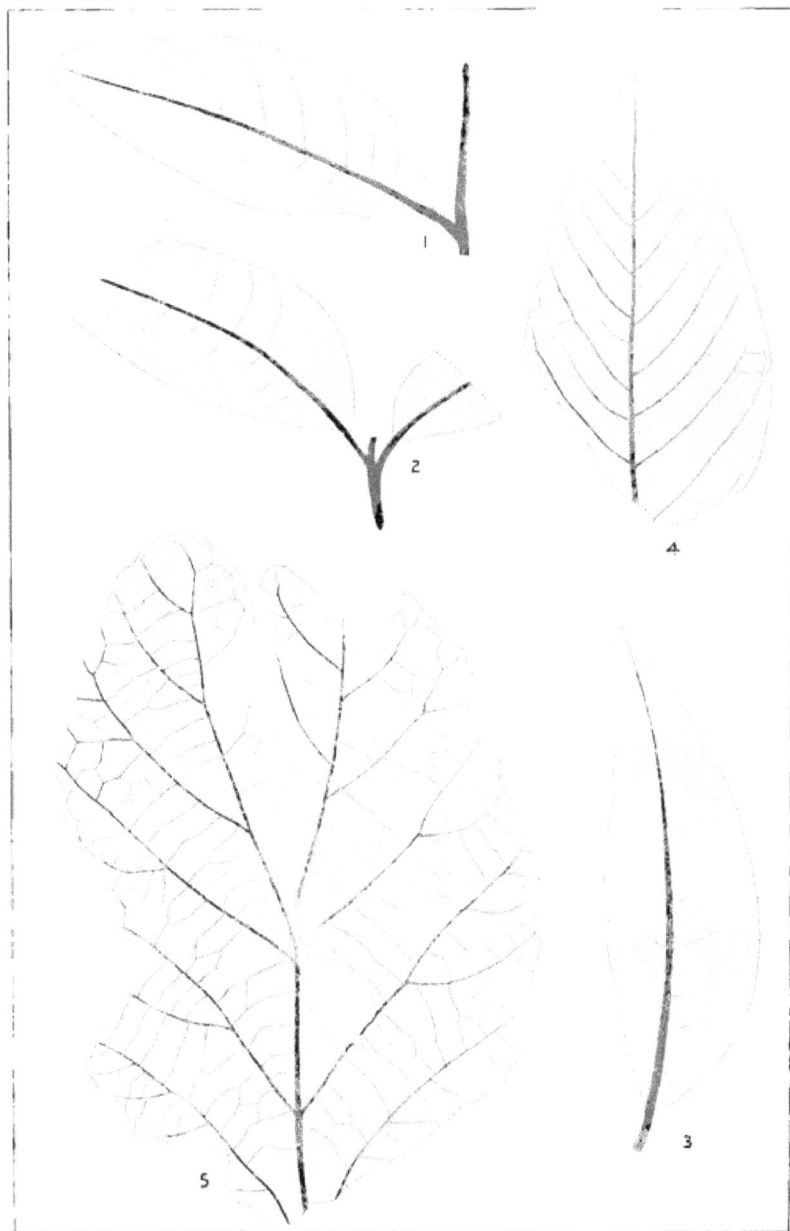

PLATE CIII.

PLATE CIII.

1

2

3

4

5

6

PLATE CIV.

PLATE CIV.

SEQUOIA MAGNIFICA

PLATE CV.

PLATE CV.

850

SEQUOIA MAGNIFICA

PLATE CVI.

PLATE CVI.

PITYOXYLON ALDERSONI.

PLATE CVII.

263

PLATE CVII.

854

PITYOXYLON AMETHYSTINUM.

PLATE CVIII.

PLATE CVIII.

856

PITYOXYLON AMETHYSTINUM

PLATE CIX.

PLATE CIX.

PLATE CX.

PLATE CX.

810

SEQUOIA MAGNIFICA.

PLATE CXI.

No. 1

PLATE CXI.

862

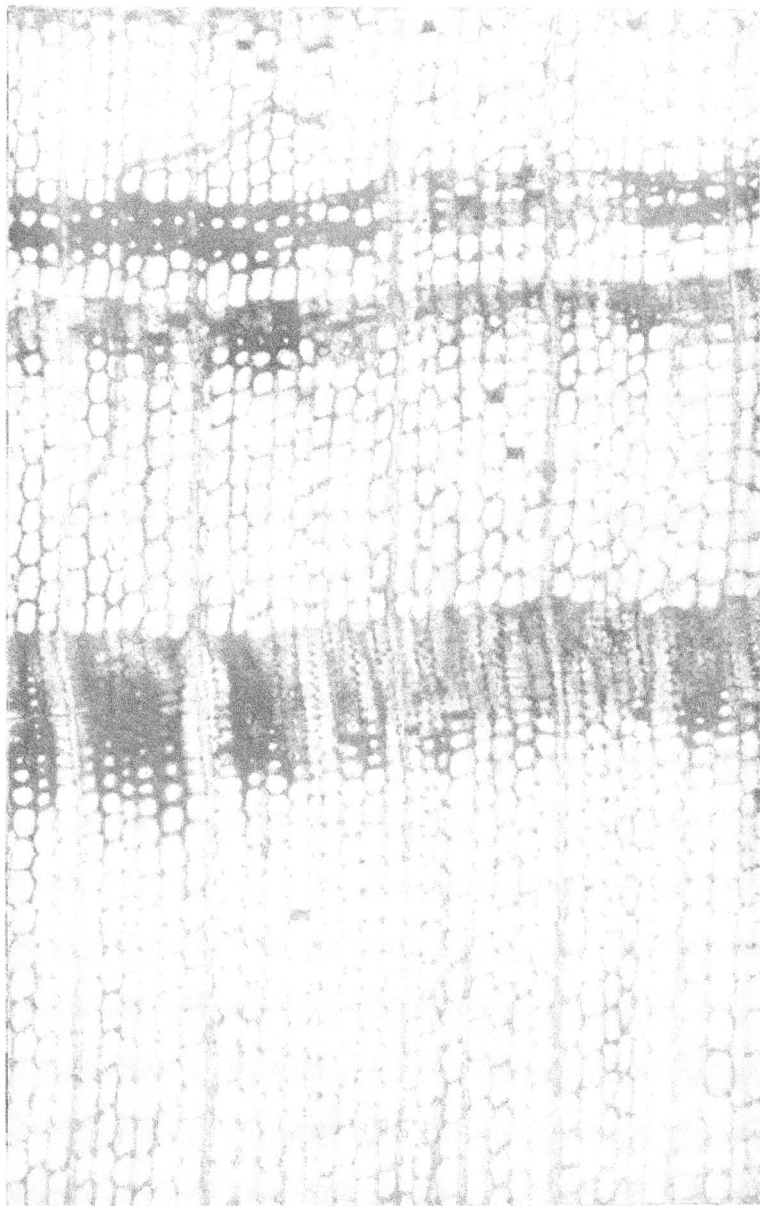

SEQUOIA MAGNIFICA

PLATE CXII.

PLATE CXII.

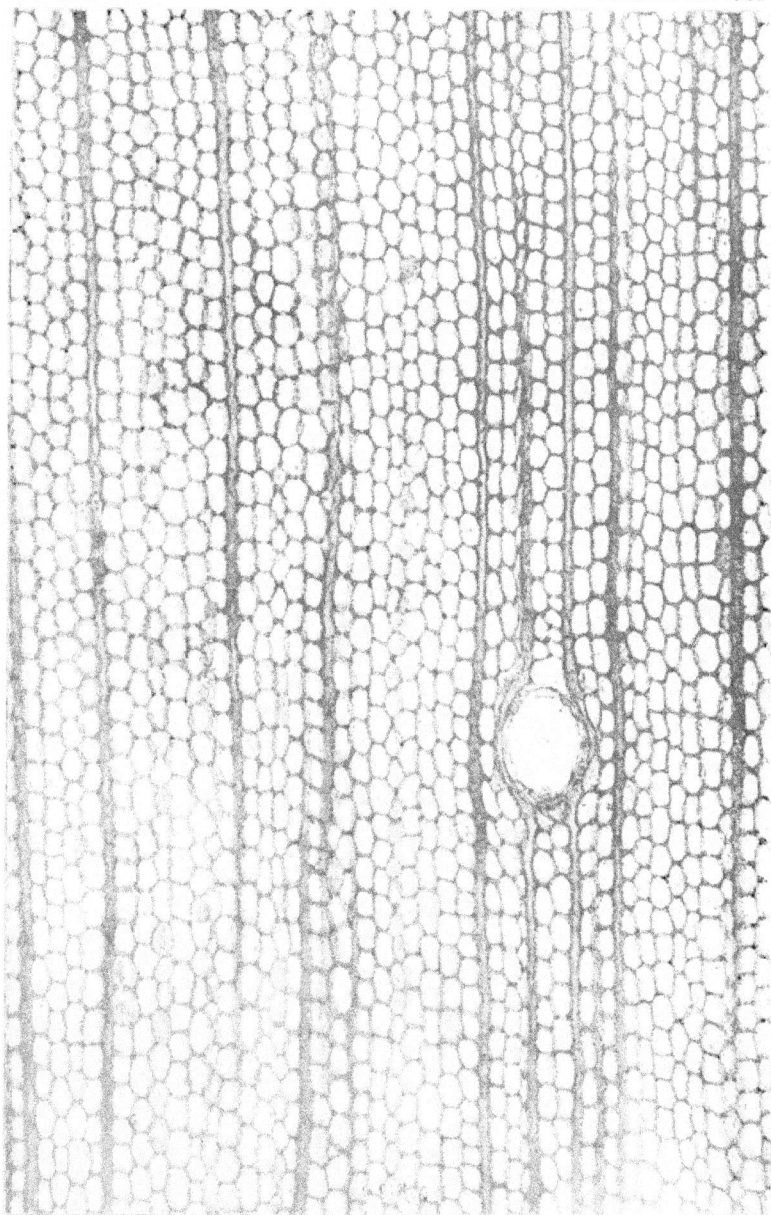

PITYOXYLON ALDERSONI.

PLATE CXIII.

PLATE CXIII.

PRIONOXYLON ALDERSONI n. sp . . . .
        Longitudinal tangential section.   Magnified 100 diameters.

        804

PITYOXYLON ALDERSONI.

PLATE CXIV.

297

PLATE CXIV.

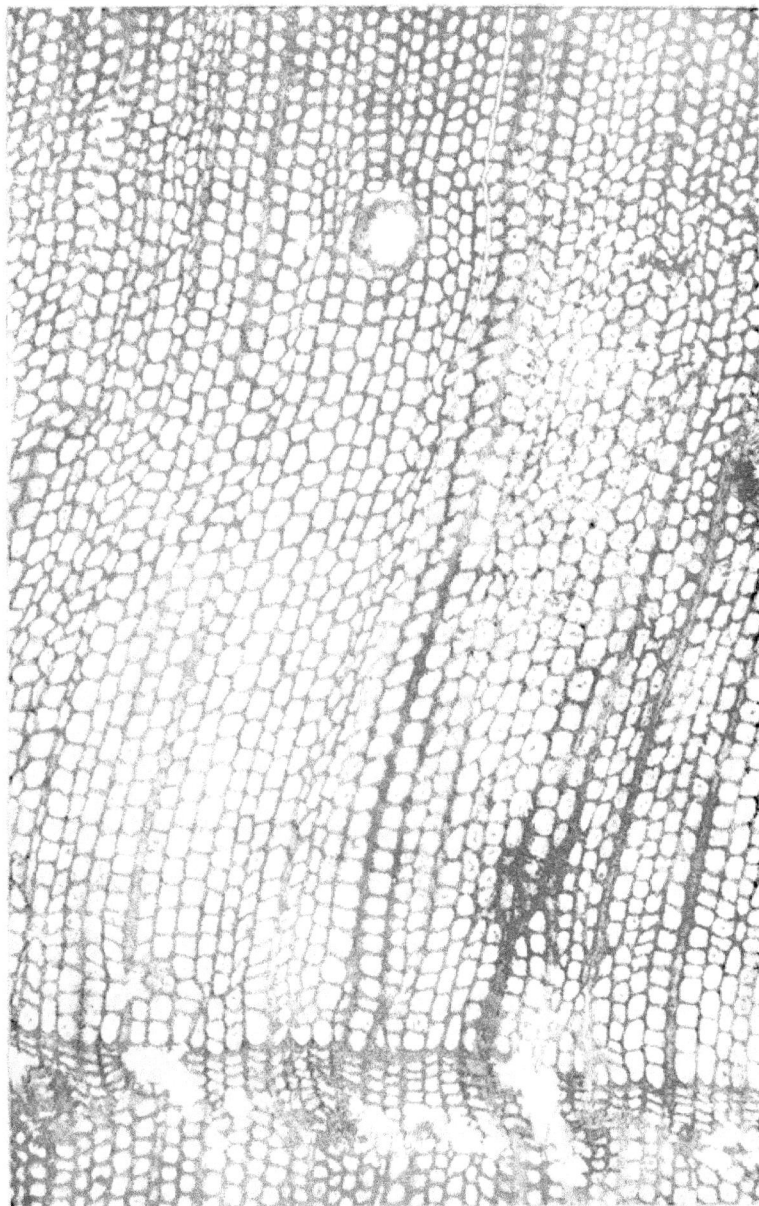

PITYOXYLON AMETHYSTINUM

PLATE CXV.

301

PLATE CXV.

PITYOXYLON AMETHYSTINUM

PLATE CXVI.

PLATE CXVI.

872

LAURINOXYLON PULCHRUM

PLATE CXVII.

PLATE CXVII.

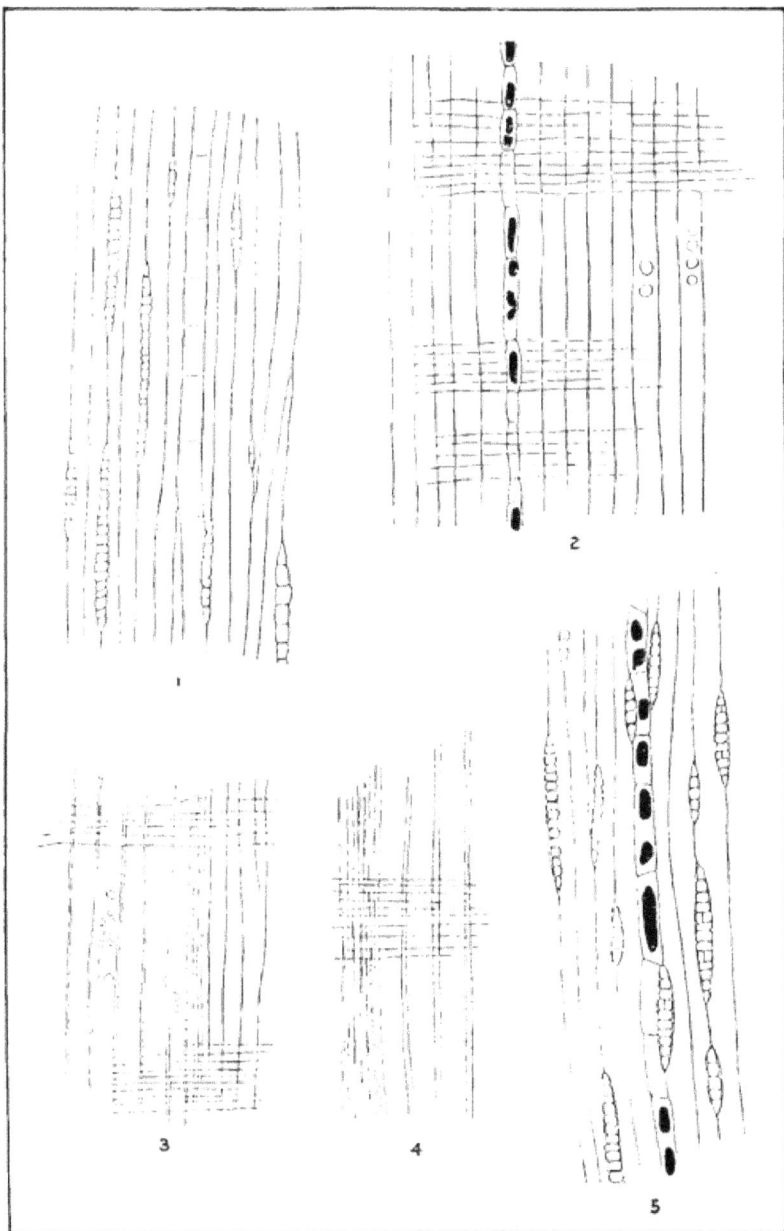

PLATE CXVIII.

PLATE CXVIII.

1. Radial section × 90 diameters. Shows wood cells with single row of pits, and spring and fall wood; also medullary rays.

2. Tangential section × 90 diameters. Shows the long wood cells and short medullary rays.

3. Radial section × 90 diameters. Shows wood cells with a single row of pits.

4. Transverse section through annual ring, × 90 diameters.

5. Transverse section × 320 diameters. Shows the fall and spring wood.

6. Transverse section × 90 diameters. Shows medullary rays of short cells, thin-walled wood cells, and series of ducts.

7. Tangential section × 90 diameters. Shows wood cells, medullary rays, and dotted ducts.

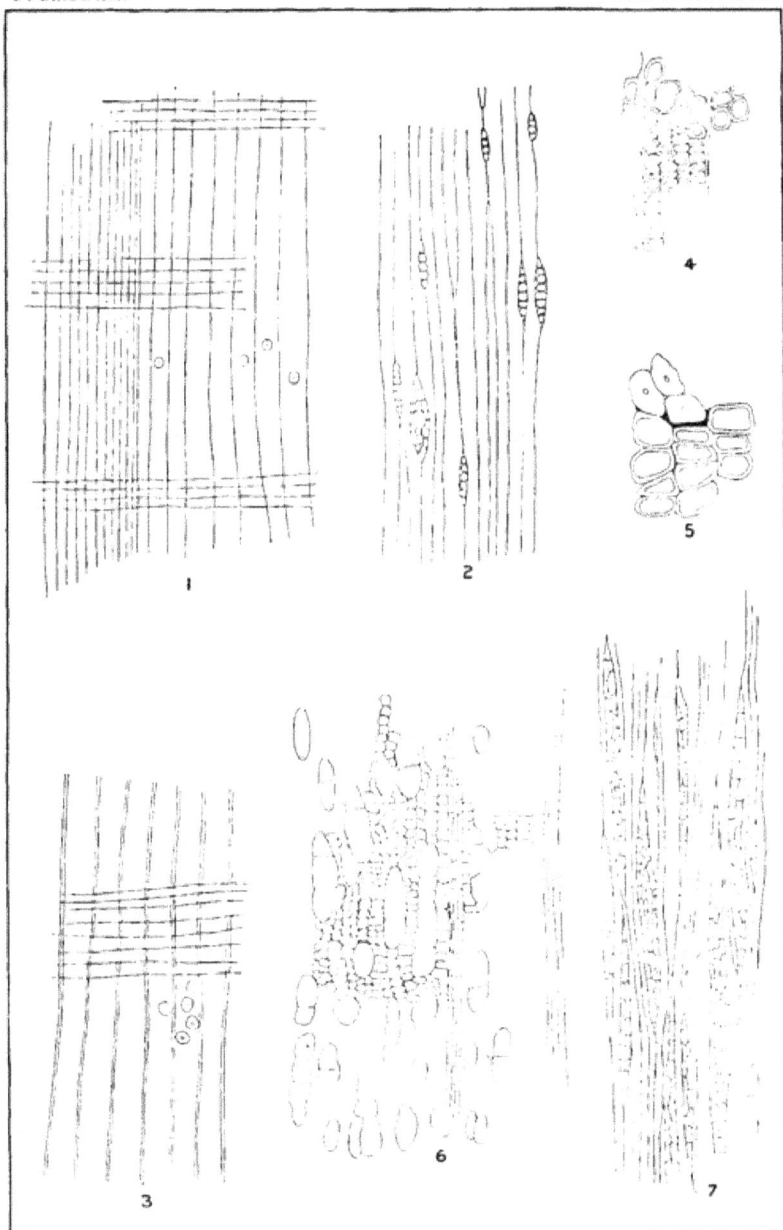

PLATE CXIX

PLATE CXIX.

1. Radial section × 90 diameters, showing wood cells, medullary rays of short cells, and large ducts.
2. Transverse section × 320 diameters, passing through an annual ring.
3. Radial section × 90 diameters, showing narrow wood cells, large plate of medullary rays of short cells, and large dotted ducts.
4. Radial section of scalariform duct × 320 diameters.
5. Radial section of dotted duct × 320 diameters.

878

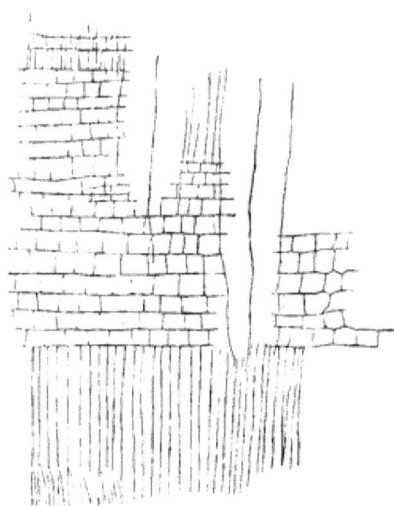

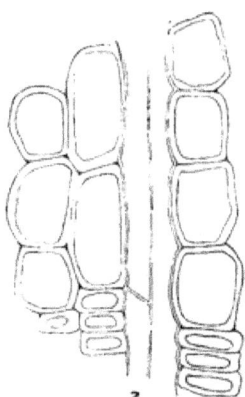

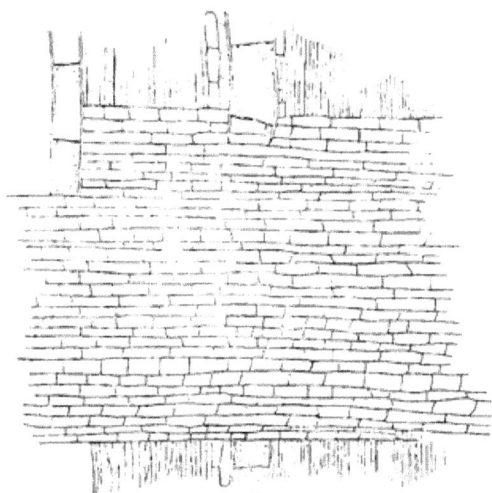

PLATE CXX.

PLATE CXX.

1. Transverse section × 90 diameters, showing wood cells of two kinds, medullary rays, and large ducts.

2. Transverse section × 320 diameters, through duct.

3. Tangential section × 90 diameters, showing wood cells, scalariform ducts, and very large medullary rays.

4. Transverse section × 90 diameters, showing short-celled medullary rays with black cell contents, wood cells in vicinity of annual ring, and numerous large ducts.

5. Radial section × 90 diameters, showing narrow wood cells, numerous ducts, and short-celled medullary rays.

880

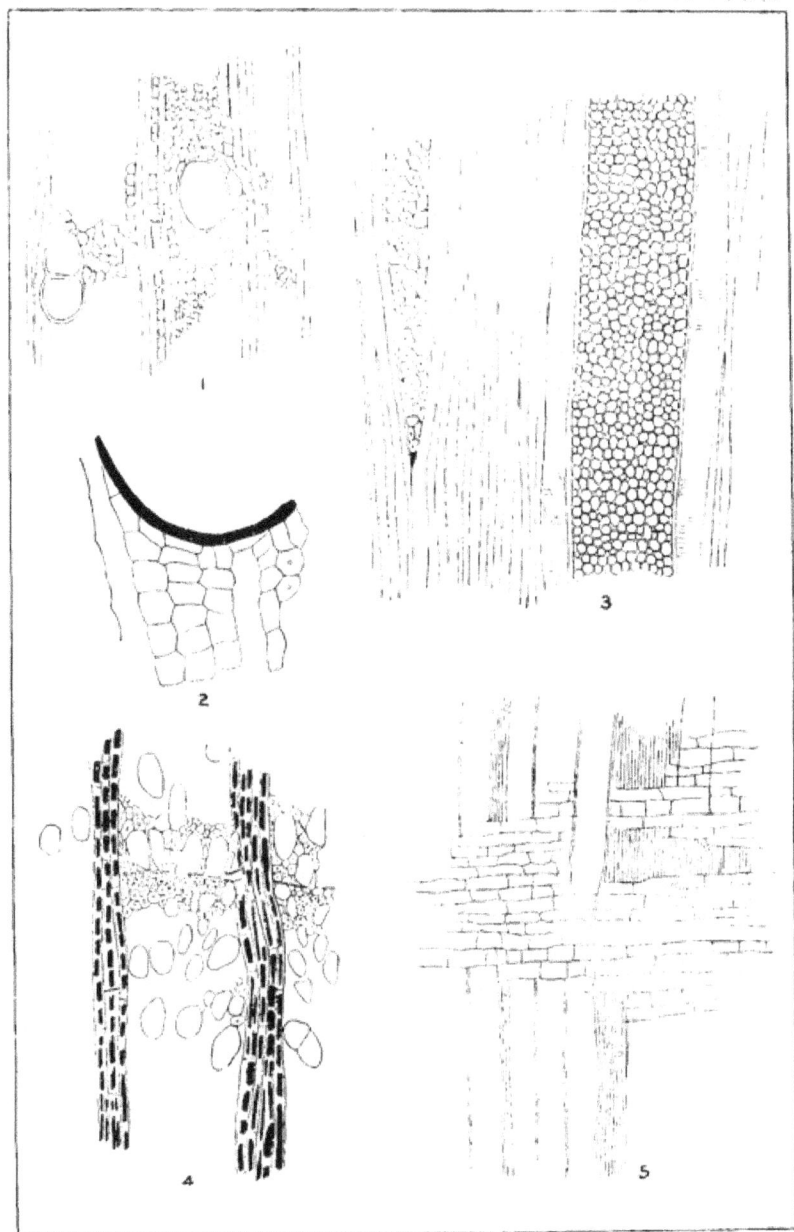

1

2

3

4

5

PLATE CXXI.

PLATE CXXI.

1. Transverse section × 80 diameters, showing thick-walled wood cells, immense ducts, and usually single celled medullary rays.

2. Tangential section × 80 diameters, showing wood cells, and large and small medullary rays.

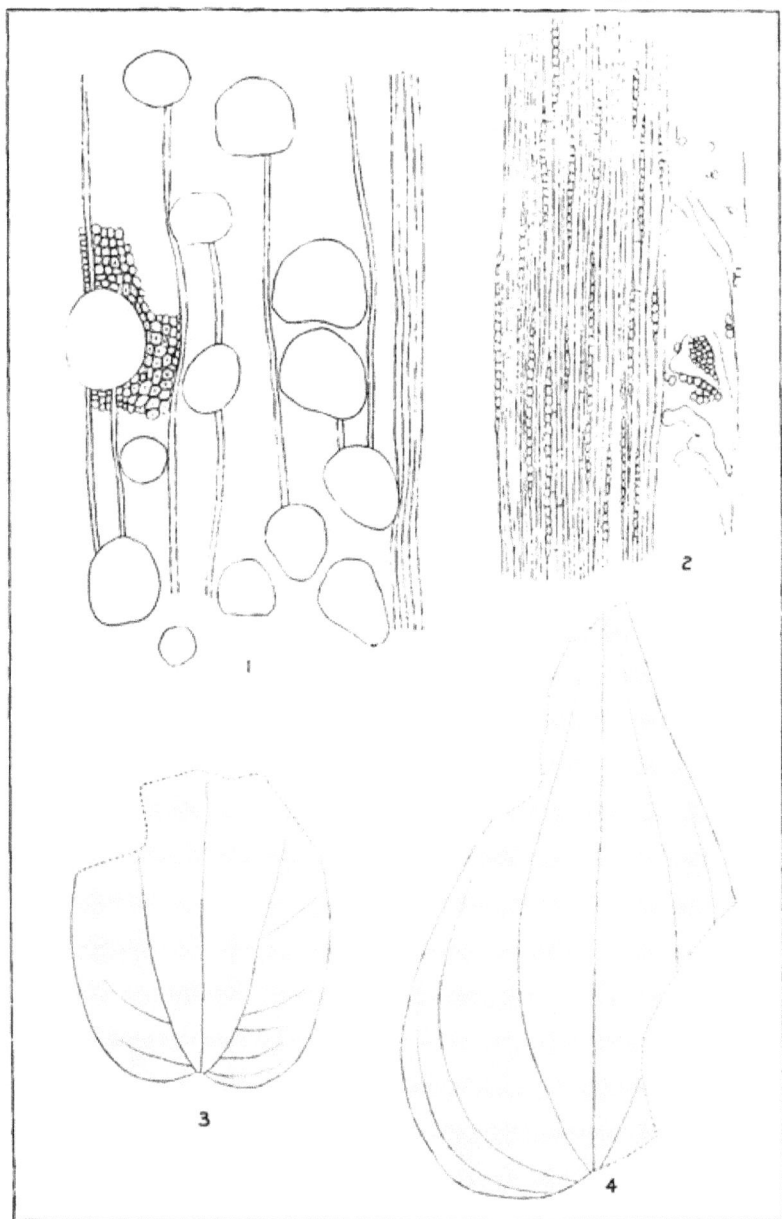

1

2

3

4

www.ingramcontent.com/pod-product-compliance
Lightning Source LLC
Chambersburg PA
CBHW021502210326

41599CB00012B/1099